The Listening Hand

Ilana Rubenfeld

How to combine
bodywork and
psychotherapy to heal
emotional pain

Foreword by Joan Borysenko, Ph.D.

PIATKUS

First published in the UK in 2001 by
Judy Piatkus (Publishers) Limited
5 Windmill Street
London W1T 2JA
e-mail: info@piatkus.co.uk

First published in the USA in 2000 by Bantam Books

A catalogue record for this book is available from the British Library

ISBN 0- 7499 2193 5

This book has been printed on paper manufactured with respect for the environment using wood from managed sustainable resources

Printed and bound in Great Britain by Biddles Ltd
www.biddles.co.uk

I dedicate this book to my father and mother,
Leopold and Bluma Churgin.
I can sense their radiant presence
and feel their unconditional love.

Contents

Foreword

I first met Ilana Rubenfeld at the Mind/Body Clinic at the New England Deaconess Hospital, Harvard Medical School, when I was directing a Mind/Body Clinic in the Division of Behavioral Medicine back in 1987. One of the early pioneers in mind/body interaction, Ilana had come to present her healing work, called the Rubenfeld Synergy Method, at a lunchtime seminar. A beautiful woman who radiated warmth, competency and humor, Ilana immediately put us at ease with a joke that broke through our armored professional veneers. Sharing laughter, we were all simply human beings. That was one of the most important lessons we learned from her. Humor heals, bonds strangers into friends, makes pain and distress bearable, and breaks down the walls that separate us from one another and from our own inner emotional lives.

Within minutes she had issued an invitation to the group for a volunteer with whom she could demonstrate her work. A psychologist by the name of Sam lay down on a massage table as fifteen or so of us looked on. A brave soul, I thought, despite Ilana's reassuring presence. Although thirteen years have passed since that demonstration, I remember it vividly. Ilana's hands were cupped so gently around Sam's head. Her face was filled with caring.

Cradled like a child, Sam began to weep. As Ilana rocked his head from side to side, placed her hands beneath his shoulders and later held his feet, she seemed to tune in to subtle bodily cues.

Speaking gently, she followed the lead of her listening hands, asking Sam what he felt. Old emotions, sadness and grief from childhood, surfaced instantly. The reality of the way his life had been constrained by the fear of abandonment brought several of us to tears. Although I knew Sam's life history, the emotions that accompanied it were much more powerful than words could possibly convey. This fact is one of the limitations of talk therapy. The body stores emotions rather than words. In order to heal, the emotions held within the body need to be acknowledged and then released. Touch is one of the best ways of facilitating this alchemical transmutation of wounds into wisdom.

With enormous grace and skill, Ilana led Sam from the prison of the past to the freedom of the present. Like a diver, she descended quickly to the depths of his soul and brought back a pearl of great price. When his body was relaxed and secure, Ilana gave words to the experience of safety. The combination of touch and the verbal offering of a new way to see the world was enormously powerful. With the precision of a laser, she changed the lenses through which Sam viewed the world, giving him an updated frame of reference. This is the work she has taught for years as the Rubenfeld Synergy Method, a powerful form of emotional and often physical healing.

"Healing" comes from the old Anglo-Saxon word *healen*, or "wholeness." Until we are aware of our emotions, wholeness is impossible, since the deepest reality of our life remains hidden. We fear the darkness, giving unacknowledged emotions power over us. When they are brought to light, accompanied by appropriate touch, powerful changes take place in the brain. Old images stored as icons in the limbic system, part of our emotional circuitry, can be erased. The result is that painful old patterns are no longer automatically engaged by present experience. When touch, emotion, new language and humor are combined, healing is even more potent. Our suffering is cut down to size.

True healers are maestros, conductors of the symphony in which the body, emotions, mind and spirit are instruments. They bring these four elemental aspects of being into harmony or synergy. It seems altogether fitting that Ilana was first trained as a conductor, for that is what she is. The human soul in its fourfold aspect is her orchestra. Healing is the emergent quality of the soul's harmony that occurs in the present moment. Feelings of peace, aliveness, curiosity, compassion, love, humor and delight naturally surface. These feelings are our birthright. They are the light of our own true nature that was hidden under the bushel of our past. We are all entitled to these feelings, not once in a while, but as the normal ground of our experience.

True healers are also teachers—educators in the true sense of the word. To educate means to lead or pull out from. They teach us to pull out hidden emotions, to develop a keener awareness of how our body and mind respond to the world. They make the unconscious conscious and show us that we can make new, more harmonious choices in our lives. We can be happy, respectful of ourselves and others, and give our gifts freely to the world. The very best healers give us the tools to lead ourselves out of darkness into light.

In this book Ilana has given us precious tools to heal ourselves. The exercises throughout the text are powerful and precise, honed through many years of healing and teaching. Each one is a jewel, capable of bringing about an "Aha." The moment of "Aha" is revelatory and life changing. In Chapter 2 Ilana lists six "Ahas" that the Rubenfeld Synergy Method brings about. In the remaining chapters you can experience the power of these revelations through the meticulously described exercises. Many of the processes are done with partners and have the added benefit of bringing the two participants into an honest, intimate, healing relationship. Doing them with a friend, spouse, lover, or in some instances with a child, can be a tremendous gift.

Ilana has trained thousands of people in the Rubenfeld Synergy Method. Her techniques are teachable, reproducible and finally available in the book you are holding in your hands. Don't just read it; commit yourself to working with it. There are many books that talk about healing. This one gives us the techniques we need to bring about true synergy. Think of yourself as the conductor of your fourfold self—body, mind, emotions and spirit. And remember to smile. Life is meant to be enjoyed, even when the orchestra is still tuning up.

Joan Borysenko, Ph.D.

The Listening Hand

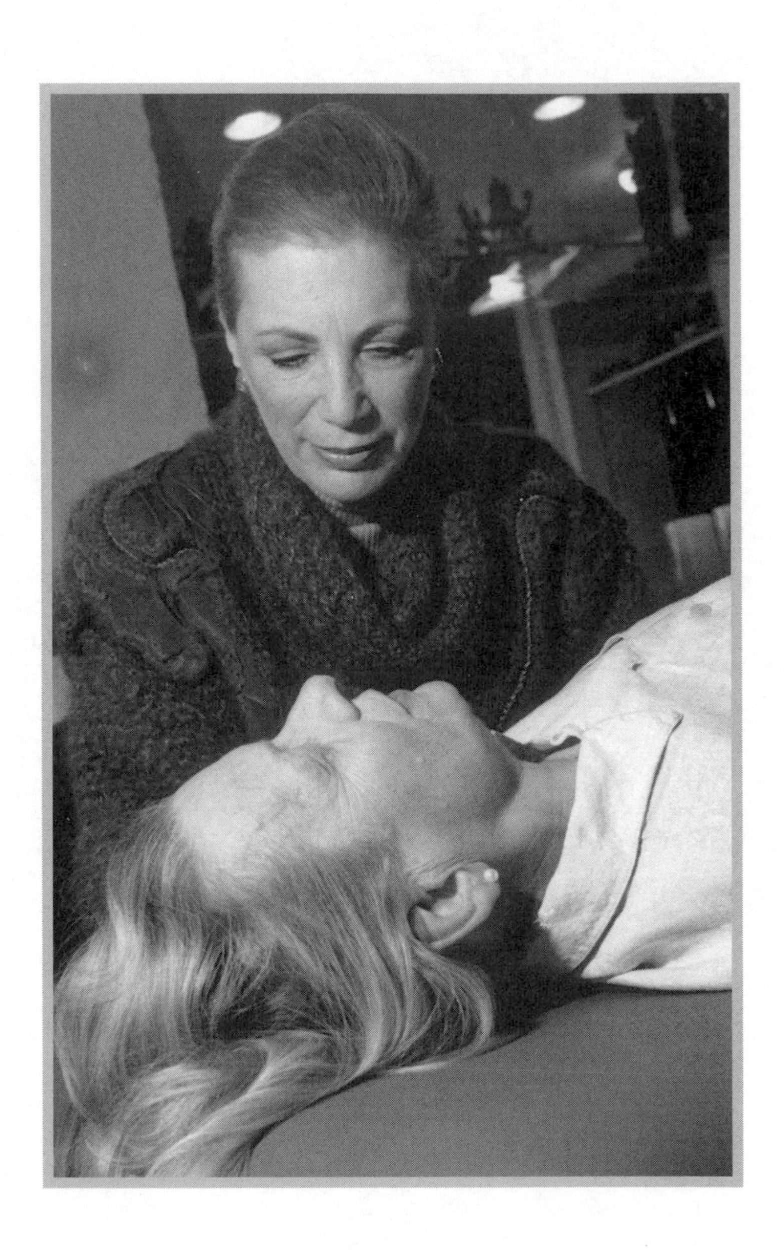

1.

How I Gave Up One Kind of Music for Another

Music saved me.

My mother, Bluma, had a beautiful voice and loved to sing. My father, Leopold, studied as a classical pianist—with the same teacher as Vladimir Horowitz. But he gave it up during the Russian Revolution when he fled Russia for Germany before going on to Palestine.

My mother's large Orthodox Jewish family also fled Russia, and landed penniless in Paris. My grandmother held them together. She was a baker, and the children sold what she baked. The family's dream was to live in America, but they were stuck in Europe. Eventually, they were allowed to emigrate to Palestine, which is where my parents met. My mother was slight, with bright blue eyes and long blond hair; she must have been a knockout. My father was brilliant, dynamic and told great jokes. I'm not surprised she was attracted to him.

Every second person contracted typhoid in those days, and Father was one of them—while my mother was pregnant with me. He spent many months in the hospital, and she lived with her family. Bluma was very anxious, and I suspect she had little time to enjoy her first pregnancy. My early memories are of my mother continually fretting.

When my father finally left the hospital, he had lost a great deal of weight and his zest for life. Somehow, though, he kept his sense of humor, and there were times when mother sang and he accompanied her. But on the whole the atmosphere in the house was dark.

Happier times were spent with my mother's sisters. With them there were laughter, games, songs—and physical affection. My first loving touch came from my aunt Hannah—Mother Earth; without her I think I would have been in serious trouble. Maybe it was with her that I first learned the importance of being held, bonding, connectedness and touch. Maybe it was because of her that I learned you did not need words to communicate.

My father was offered a good job in Paris, but our exhilaration was short-lived, for he got seriously sick again and by the time he was well enough the position was gone. He was promised a job in the U.S.A. and decided to go. Bluma did not want to leave her clan, but her father told her that a good wife always followed her husband. So she and I sailed for America. We were lucky. We were on the last boat to leave Palestine for New York until after the war. A few weeks after we sailed, Tel Aviv was bombed. The house and dreams of my childhood and those of all our friends were blown to pieces. I was not yet six years old.

When I came to New York, I had no one to turn to. My father was struggling to get work, so I barely saw him at all. My mother was in a state of shock and deeply depressed. I didn't understand or speak English, and there was no one at home to help me.

School, of course, was taught in English. For months I sat in the first-grade classroom, not understanding a word. It was frightening, frustrating and mortifying. This experience shaped my life and later my career. I had to decipher the true meaning of what was being said from body language—posture, gestures, movements, facial expressions and voice tone. This skill allowed me to see behind words and hear their true meaning. It became clear that contradicting messages were being given when the words said one thing, the body quite another.

From home to school and back was a harrowing daily journey through a religious Irish-Catholic neighborhood. I was pelted with rock-filled snowballs while my tormentors shouted "Christ-killer." What a sad and painful time it was for a little silent Jewish girl plodding to and from a school where she understood no one to a home where no one understood her.

I pleaded with my mother to explain what a "Christ-killer" was. Somehow

I was to blame for His death. My mother, flustered as well, tried her best to assure me I was not His murderer. But to no avail.

The torment and the emotional wounding continued. I missed the Mediterranean Sea, the trees of Tel Aviv and, most of all, my extended family. My mother cried, and I remember fighting not to imitate her. I was a displaced person. Lost.

Then someone made a donation to our family worth more to me than the crown jewels of Russia.

A piano.

If nothing else, the piano brought me and my father together. He played and I listened. How I treasured the sound! He played Russian music and Chopin. As a child, I loved Snow White and Donald Duck, but by the time I was a teenager it was the mazurkas and *The Rite of Spring.*

My parents sent me to a Catholic summer camp to get me out of New York, which they were sure was in danger of imminent attack. It was there that I continued to learn English by having to say my Hail Mary every morning and the Lord's Prayer at night.

The counselors quickly discovered I had a good ear and a nice voice. I would run to the camp piano and play for hours. I couldn't read music but I could compose it, and in time I composed the music for the camp play and directed it. At age eight I was already a conductor!

Every Russian mother believes that little girls should study ballet, so mine enrolled me in a class on Saturday mornings. Yolishka, a traditional Russian ballet teacher, hit the floor and her pupils with a long wooden stick. There was always a reward afterward: my mother would take me to a Hungarian restaurant for stuffed cabbage, and then to a matinee of ballet at the New York City Center. It was our ritual, our closest contact.

My father frowned on music lessons because they were impractical, and he insisted I learn how to type. He repeatedly said that there would always be a secure job—something he never really had—for a typist.

News of relations killed in Europe trickled in at first, then began to arrive more frequently. I was terribly upset by what I heard, and pictures I saw. The time of the Holocaust haunted me. Years later, as a beginning therapist, I had trouble separating from my own angst when I treated camp survivors.

I was constantly badgering my parents with questions.

For my father:

How could the Germans produce Beethoven yet kill all those innocent people?

How could God let little children be killed?

How could there be concentration camps if there was a God?

For my mother:

How come *you* are so unhappy?

My parents couldn't answer, and my frustration with them grew. Then, when I was twelve, my sister Lydia was born. My mother immediately went into a profound depression and disappeared from home. It was not until much later that I learned she had been hospitalized.

I was left to take care of my sister, and became her little mother. When my mother returned, I took care of her, too. Now I was both my parents' mother and my sister's mother—as well as a student at junior high school.

It was at the High School of Music and Art that I began to develop a love of conducting. I also learned to play the viola, a tough instrument for a young girl, requiring strength and a strange sitting position. I began developing the back problems that had a great influence on my exchanging a career in music for a career in bodywork.

During these teenage years, I also joined a Jewish scout organization that believed American Jews should return to Israel to become farmers. One day the conductor of the scout choir got sick. "You go to Music and Art," I was told. "*You* be the conductor." It was both a frightening command and music to my ears.

Before starting college, I *did* go to Israel to work on a farm and study. I wanted to take care of the earth, making the world a better place to live. (I also conducted choral concerts of simple folk music.) It was an attempt to find my roots. I came back feeling more whole and more grounded than at any previous time in my life.

When I returned to the States, I went first to the Manhattan School of Music, but I transferred to Juilliard because there I could major in conducting.

One voice at a time, I told myself. Soprano, alto, tenor, bass. Strings, woodwinds, tympany, brass, etc. Then put the parts together.

Soon I was able to hear the whole from the sum of its parts and then I learned to transmit this vision to the players, not through verbal commands but by using my face, my body and especially my hands. I could *cradle* the music, hold it in my hands and hold it out to the players. My hands could both listen and speak.

Through conducting, I came to understand the complexities of listening, talking and moving simultaneously. Like music, healing involves the capacity to listen to others and hear their inner song. Hearing silence as well as sounds is part of music, so the healer learns to listen to both sounds and silences. Impulses, needs, emotions and feelings in people are expressed through sounds and silences, words and wordless movements of the body. The healer listens to all these variations and helps the client achieve harmony.

This notion of emotional orchestration coalesced for me when I conducted Vivaldi's *Gloria* Mass. Standing at center stage, I became a musical and spiritual bridge between the singers in front of me and the audience behind. As we began the second movement, "*et in terra pax homnibus*" ("and on this earth, let us have peace"), a life-shaping event occurred. The music became extremely intense, and I left my physical body and in a spiral movement floated up to the ceiling. Surprised, I looked down. I saw my parents, my teachers, my friends, and I watched myself conducting. I was both inside and outside the musical notes on the page, experiencing the sounds they made, becoming one with the energy of the music. I was utterly at peace; in heaven.

Suddenly, the cello section came in a measure too soon, and I was hurtled back into my body to rectify the mistake. As the last chords ended the magnificent "Sanctus," I turned around to face the audience and bowed. I saw their hands applauding, and yet there was no sound. Backstage, people talked, but I didn't hear them. I smiled. Everyone looked beautiful. I was still at peace.

It was several days before I realized that I had undergone a deeply religious out-of-body experience and had altered my state of consciousness. Leaving my body both frightened me and inspired me. It was a spiritual, peaceful and egoless state. As the conductor/bridge, I was making the music happen, but also stepping aside from it, letting it happen on its own, listening to it happen, *hearing* it happen. I became an insignificant part of the music's energy, and also essential to it. I had re-created it, but it had also re-created me.

This same state of ego and egolessness, of insignificance and essence, is true in regard to cosmic energy, of which music is a magnificent expression, for we are all insignificant, essential parts of the cosmic sphere, part of the ascending spiral. And it is true in regard to my therapeutic work with people.

When I'm with clients, I feel that I am in that egoless state. I am still Ilana, yet different: nonjudgmental, totally accepting, loving. I can be both a partner in the therapeutic experience and step aside from it. You sing your own song. I am a vital guide to help you hear and understand it, and to accept your body as your musical score.

The tools I use for my therapeutic work—to let you hear your authentic self—are my brain and my hands. For as I learned early on, and shall describe in detail in the following chapters, emotions and feelings are embedded both in our body and in our brain, and to reach them, we need touch—nonverbal communication—and talk. Through touch you can understand the body's message, and through talk you can understand the brain's, even if the message is "hiding" in the unconscious. Body and brain are equally revelatory. Together they send meaning that through either alone is incomplete.

Fritz Perls, the co-founder of Gestalt therapy, insisted that I was a composer as well as a conductor. In a way, he was right. A conductor takes other people's music and brings life to it. Composers create unique forms of music. I composed the Rubenfeld Synergy Method, which has a unique music of its own.

Conducting for many hours a day challenged me physically. At that time, at Juilliard, no one had taught me how to use my body efficiently. After graduating, I accepted an assistant conducting job with Leopold Stokowski, and another position as a conductor in the 92nd St. YM-YWHA music department. The pain which had begun when I was learning the viola now returned with redoubled force. As I was preparing for a concert, my back and shoulders went into spasm. I couldn't lift my right arm—for a conductor, a disaster.

Various doctors prescribed medications that gave me only temporary relief. I could get through a concert, but the pain quickly returned. Desperate, I began looking for other ways of treating my condition. A musician friend suggested that I try the F. M. Alexander Technique, a method designed to teach balanced and efficient posture. The therapist he recommended was a woman named Judy Leibowitz, the only person teaching the Alexander Technique at the time. I made an appointment—and, in the very first session, found my true vocation.

The revelations I experienced in Judy's office were as powerful in their way as those I'd had when I was conducting the Vivaldi Mass. They led me to a series of revelations (I call them "Ahas") that have become the foundation of the Rubenfeld Synergy Method. I'll describe them in the next chapter, and you'll see them in action throughout the book.

For the purposes of this brief autobiography, however, the important factor was that by studying and applying the Alexander Technique I was able to resume conducting without debilitating discomfort.

Although the pain had disappeared, I continued working with Judy. Now it was not pain that prevented my conducting, but the fact that I was a woman. I had made a promising start, conducting at Carnegie Hall and at Town Hall. But time and again I was turned down for jobs because I was female. If I had a penis, I thought, I'd be assisting Pierre Monteux, who refused even to accept me in his conducting class. "I don't teach women conducting!" he wrote in response to my application.

Gradually, as my interest in the Alexander Technique increased, the pain of rejection diminished. Judy encouraged me to become a teacher, and for a time I worked actively in both fields, music and bodywork.

Although I wasn't a psychotherapist yet, I was convinced that posture and touch played an important role in accessing emotions. In the process of one's changing posture, emotions were being shunted aside by the body, hidden and not dealt with. Yet there were emotions, I was convinced, frozen along with the memories that elicited them in the tissue of the body as well as in the mind.

In the mid-1960s, I met Fritz and Laura Perls. Fritz was teaching a small group at the Esalen Institute, and he became a mentor to me. It was through his teaching that I added more psychotherapy to my own therapeutic theories. He invited me to use touch with his clients while he talked to them. I soon reconfirmed that there was a subtle and clear muscular response to every thought and emotion that people felt. And I learned that touch could open gateways to the mind.

It was at the Esalen Institute that Moshe Feldenkrais began to train and teach a few of us. For the following eight years I trained in his two-tiered method of Functional Integration and Awareness Through Movement, while I continued to follow my own explorations.

Both the Alexander Technique and Feldenkrais's method taught people how to release tense habitual patterns held in the body. Neither, however, addressed the emotions behind those patterns, or their often vivid expressions when the patterns were unlocked. Conventional psychotherapy relied on talking through problems, and the bodywork addressed their physical aspects. To me, these disparate approaches were fragmented, denying the unity of the body, mind and emotions.

In the 1930s, the controversial psychiatrist Wilhelm Reich had espoused the same concept of combining bodywork with psychoanalysis. He believed that emotions are held in the body, "blocked" there by "armor" which prevents their release. He emphasized that all living creatures have energy and that when this energy is blocked, it causes the life force to diminish, resulting in characterological defenses that rob the client of freedom and authenticity. Reich used intense, deep-pressured touch to "break" that armor. I found that using light, nonintrusive touch "melted" the armor. The goal was the same: the release of emotion and feeling to allow the client to become his/her full energetic self.

Putting mind and body therapy together was as exciting as making music—quite literally, it was the discovery of harmony of a different sort.

I realized that for people to understand their emotional state, they would have to learn how to listen to their bodies. While observation and talk are valid tools, one of the most powerful ways to heal is through the use of touch. Often, emotions are beyond words; they are not linear or rational. Hands are the most sensitive receptors of the subtleties of sensation. I learned—and you can learn—to develop "listening hands" in order to "hear" changes in the body.

Listening touch became my instrument for tuning into all levels of the psyche. Touch and talk became an essential duet; here were two great healing forces that when synergized were greater than either alone.

The medical meaning of the word "synergy" is "a remedy acting similarly to another remedy and increasing its efficiency when combined with it." It was Buckminster Fuller who suggested that it was appropriate to call my work Rubenfeld Synergy and to call myself a Synergist. Semantics, perhaps, but it expresses the basis of this work, because I strongly believe that everything—mind and body, thoughts, emotions and feelings, bloodstream and brain, soul and spirit—must be in a state of synergy if we ourselves are to become whole.

The word slowly spread that mine was an effective approach. People with many different needs and backgrounds began to come to my office. I worked with world-famous performers, other therapists, the wife of a president—and many other people who longed to be balanced and well.

In these beginning years, I was frequently approached to write a book about this new approach. However, I was still refining the theories and philosophies, and developing exercises that anyone could do. Life force and energetic concepts were difficult to translate onto paper. Only recently have scientists begun to document the interaction of mind, body and emotions.

At the beginning of the twenty-first century, the way to healing has become less fractionalized; bringing the body, mind and spirit into a congruent whole is more common than it was even ten years ago. I write this book now because I can illustrate what happens in the healing process, and by so doing show you the way to self-healing. I will teach you about the way your body holds emotion and the methods for releasing it; about your energy field, which is the basis for therapeutic touch; about the use of gentle, healing touch; and about the all-important factor of awareness, which is the first step toward change. This book will provide you with ample instruction and exercises for you to experience change and adopt it into your life.

It is meant to inspire. It is meant to help you understand your emotional body, and to help others understand theirs.

2.

The Rubenfeld Synergy Method®: The Six "Ahas"

Come back with me to my first encounter with Judy Leibowitz. She escorted me into a small room with a mirror and invited me to lie down fully clothed on a padded table. Then she did something no one had ever done: she touched the sides of my head very gently and asked me in a soft voice to relax. I followed her instructions—or so I thought, for I was soon surprised to hear her repeat her request.

"I *am* relaxed," I told her, talking through clenched teeth, tight jaw and taut neck.

"Not really. Try to sense your neck and head and what you're doing with them."

My head and neck were rigid with tension, as I discovered when Judy tried to gently move them. I'd only *thought* I had relaxed. I was amazed to discover that I was not doing what I'd sincerely believed I was.

I had come to my first "Aha."

Aha #1: You're not always doing what you think you're doing.

This may sound simple—but it is an extraordinarily important insight. You may think that your mind and body are communicating, that your body is following your mental instructions, when in reality your mind is telling your body to do one thing and it is doing just the opposite.

While touching my neck, Judy repeated instructions my brain didn't understand. However, *my body understood the message in her touch,* which cajoled my head and neck muscles to soften, allowing me to experience a new freedom of movement. Another "Aha" quickly followed:

Aha #2: Touch is a powerful means of communication.

Judy's touch succeeded in communicating with me more clearly than her words, conveying messages of nonviolence and safety to ease my severe tensions. I had never experienced this kind of touch before, I realized—touch that talked to me. It was only later that I understood this is the way babies experience safety in the arms of their parents.

After a while Judy stopped her touching. I sat up and got off the table.

"What do you do?" Judy asked as I stood there.

I was startled and surprised. None of the doctors I had consulted had asked me this question. "I'm a musician. I conduct."

"Show me," she said cheerfully.

I began to conduct an imaginary orchestra.

"Of course you have back and shoulder spasms. Your neck is sticking way out, your right shoulder is slumped and your left ankle is totally out of alignment."

"What does my left ankle have to do with my neck and right shoulder?" I asked.

Judy explained that what happens to one part of the body affects and ripples throughout all other parts as well. In my case, an old ankle injury was manifesting itself and having an impact on my back and shoulders. Continued hours of conducting had exaggerated the already poor posture of my upper body.

My third "Aha" emerged as a bright light.

Aha #3: The body is an interconnected system.

My first session was an astonishing success. Feeling taller, balanced, lighter and more relaxed, I rushed to see the friend who had recommended Judy to me.

"How was it?" she asked.

"Wonderful!"

"What happened?"

I stared at her. "I don't know," I muttered.

As I struggled to explain, the fourth "Aha" presented itself:

Aha #4: To have an experience, you don't have to know the reasons why.

I had just had a life-changing experience. I couldn't explain it to my friend, couldn't "justify" it to myself and did not cognitively understand what had happened. All I knew was that something profound had taken place, and that if I had tried to stop and analyze it—if, indeed, I had paused during my session to try to formulate the "Ahas" into words—I would have weakened the impact of those precious moments.

Over the next few months, my spasms dissolved and I was able to conduct without pain. I never missed my weekly appointment, but during one lesson with Judy, a surprising event happened: I began sobbing from a place deep within me. Her gentle touch had built such trust that I was able to safely contact and express sadness in her presence, sadness I had denied even to myself. After several of these emotional episodes, I realized the fifth significant "Aha."

Aha #5: Emotions reside and are held in the body.

At first, my recognition of this truth was more physical than mental. When Judy touched me, I often cried. A vast store of emotions had awakened and was being released. Somewhat at a loss about what to do with these emotional episodes, Judy finally suggested I have a consultation with the analyst she went to. She explained that her teaching was about the efficient use and alignment of the body, and that she was not qualified to process, interpret or talk about the feelings and memories that emerged during our lessons.

Full of emotional pain, I promptly made an appointment with her psychiatrist. He invited me to lie down on his couch, free-associate and recall the emotions I had just experienced with Judy. I could not describe them—nor experience them with the same intensity. I suggested that perhaps touch had had something to do with the intensity of the feelings, and I asked him if he would try to use this approach as well. He recoiled, telling me he would never use touch in the psychotherapeutic process.

Thus the next several years were divided between Judy, who touched but would not talk, and the analyst, who would talk but would not touch. I longed to have a single professional do both, and when I couldn't find one, the sixth "Aha" presented itself.

Aha #6: Why couldn't someone do both—talk and touch?

I knew within myself that I was that "someone." Without a model for working with the body, mind and emotions simultaneously, I began to experiment with verbal psychological interventions while using the Alexander Technique. I became the bridge between two disciplines—bodywork and psychotherapy. And the Rubenfeld Synergy Method was born.

The memory of one of my early clients is still vivid today.

Susan came to me because of severe shoulder pain accompanied by unremitting anxiety. By the end of our first session, using touch and postural self-awareness, I was able to get her to release her rigid shoulders, but when she arrived for the following session she was as hunched up and tight as before. Obviously, touch and postural self-awareness had been only a temporary remedy.

Several weeks later, in the midst of a session, she suddenly began to cry in a very high-pitched, frightened voice. Her body became rigid and she bowed her head in shame.

"How old do you feel right now?" I asked, recognizing the significance of the moment.

"Two years old," she whispered.

While my hands softly touched her upper back and right shoulder, I asked her to close her eyes and go back to that age. She suddenly shuddered and squirmed in her chair, then brought her knees up to her chest so tightly she resembled a small ball. She peeked out briefly to check that I was still there. Through my hands and voice I assured her I would not leave her. A distant memory surfaced. Her mother was tying her hands with brightly colored ribbons to the side bars of her crib.

Her story came out haltingly. Each time baby Susan would touch her genitals, her mother would push her hands away and tie them to the crib. This memory was long repressed, but her position—and the shame—had remained frozen in her body.

In the sessions that followed, she was able to release her shoulders and keep them relaxed. Now her arms moved more freely, allowing her hands to come closer to her body—touching herself. This new body position scared her at first, and she often went back to her old frozen posture. But after some months of working through this emotional trauma, she was able to reclaim her sexual feelings without fear. Susan even entered a fulfilling relationship, her first, with a man. Eventually, she was able to forgive her disturbed mother. She had finally integrated her relaxed shoulders and expanded open chest into her present life and relationships. And I had been able to talk to her and elicit talk from her, as well as to use touch to reeducate her body.

After Susan, I found it difficult to practice the Alexander Technique by itself. I was no longer satisfied with changing only my clients' postures. I burned to know their histories and what had created their dysfunctions in the first place. Listening to their physical tensions was a gateway to a larger story that began to unfold beneath my hands.

The "Ahas" were the first steps toward developing a full therapeutic program, but they do not constitute the entire theoretical construct behind Rubenfeld Synergy. This developed slowly as my number of clients grew and I saw the efficacy of my methods in "real life" and, later, when I gave up much of my private practice to devote myself to training others in my method.

In 1987, I "dared" to publish some of my theories in several journals. Later, I expanded these concepts and included them in a book for professionals on body-oriented psychotherapy. Here, for the first time, I present them to the general reader.

The following principles, philosophies and theoretical foundations, all of which will be treated in depth in the chapters to come, guide Rubenfeld Synergy sessions:

1. Each individual is unique.

We approach each client as a separate human being, recognizing his or her uniqueness; there are no characterological maps or templates, as exist in some other systems, which purport to correlate emotional behavior to various areas of the body (back pain = anger, chest constriction = sadness, etc.). Synergists approach each session with each person without a predetermined agenda, choosing from options of touch, verbal interaction, imagination and movement to support the client's unique path to growth and change.

2. The body, mind, emotions and spirit are part of a dynamically interrelated system.

This is the principle derived from "Ahas" number 3 and 5, and I will expand on it as the book goes on. The fact that body and mind are interconnected—a radical concept thirty years ago but commonplace now—lies at the base of many therapies. The *emotional* and *spiritual* elements are later developments and were unique to Rubenfeld Synergy, though others have now adopted them. Essentially, every time a change is introduced at one level of a person's being, it has a ripple effect throughout the entire physical, emotional, mental and spiritual system, changing the equilibrium of the whole person. Movement and posture become expressive of our entire relationship with ourselves—and with the rest of the universe.

3. Awareness is the first key to change.

Each of us has physical and emotional habit patterns. But we may not recognize them and how they affect our lives, because we learned them unconsciously. To change these habit patterns, we must become aware of them. Through movement, touch, verbal intervention and creative experimentation, we can heighten awareness of our habit patterns and thus begin to make different life choices.

For example: David stood with his right shoulder frozen much higher than his left. He complained to me of chronic neck pains. Obviously, he was not aware that his shoulders were very uneven and that they were producing the pains. Slowly, he came to realize that one side of his body was much different from the other. By the end of the fourth session his shoulders were even. I asked him how he experienced himself. With his eyes closed, he reported that now his *left* shoulder felt much higher. When I asked him to look into the mirror, he was astounded. There he was with a pair of even shoulders, yet with his eyes closed, his inner experience was the opposite. Reality and his old perceptual habit were at odds. We had to practice this perceptual awareness and distinction many times before he was able to accept and integrate them. He would need further work to recognize the emotional content his shoulders had buried.

Your patterns won't be David's; you have your own. And once you're aware of them, you're on the road toward acceptance and change.

4. Change occurs in the present moment.

You may experience your memories of the past and fantasize about the future, but change itself can occur only in the present. When memories of past painful experiences emerge in a Rubenfeld Synergy session, clients can relive the experience in real time through imagination and visualization. After a while, they can rescript these remembered events in their imaginations *and in their bodies* and look at them from another perspective—quite a difference from looking at them through talk and interpretation alone. Thus they can resolve unfinished emotional business and integrate new insights into the present. By looking (both physically and intellectually) in the present into the cloudy past, they can cope with hitherto denied or unresolved anguish. And you can too.

5. The ultimate responsibility for change rests with you.

Awareness is essential to change, but it is not enough. The old joke is in fact true: "How many psychotherapists does it take to change a lightbulb?" (Pause.) "One . . . and only if the bulb wants to change." You, too, must *want* to change; no one can force you to. Indeed, sometimes the very resistance to change or the denial that change is necessary is what holds people together, even if it is the basis for their dysfunction. The only thing a Synergist can do is help them recognize the dysfunction—emotionally and physically—and slowly guide them to a new behavior that can replace the habits of the past. The process can be slow—as it should be. When Synergists touch a tight holding area, they may palpate it slowly and gently, all the while verbally exploring various themes. If the area does not soften or release, however, it's a clue of possible resistance. Synergists, respecting the resistance, do not force an entry but return there at another time. (This is different from Rolfing, say, and other kinds of bodywork.) When change and release *do* occur, they are initiated by the client *from the inside out,* and not forced or pushed by the Synergists *from the outside in.* At best, the Synergist's touch and verbal processing serve as a catalyst for the client's change.

6. People have a natural capacity for self-healing and self-regulation.

The client's innate healing ability already exists, waiting to be actualized. A basic tenet of Gestalt psychology is that every organism has the ability to attain the optimum healing balance within itself and its environment. A 1995 Harvard study shows, for example, that the simple act of remembering how it feels to take care of someone, or to be cared for by a loved one, can drive up—and sustain for an hour—the body's level of the antibody immunoglobulin A! Other studies show a link between loneliness and lowered cellular immunity.

And Dr. Candace Pert, in her exploration of neuropeptides (the hormones that govern communication between the brain and body cells), found that the limbic system—the part of the brain responsible for moods and emotions—has forty times more neuropeptide receptors than other parts of the brain. There is no question now that body and mind are chemically and physiologically linked; scientific proof abounds. Work that involves both the body and the emotions fosters self-healing by giving clients access to the brain center where their deepest emotional experiences are encoded.

7. The body's life force and energy field can be sensed.

There are many ancient energy systems that became known in the West only in the twentieth century. The body's energy has many names. "Chakra" is a Sanskrit word that describes circles of energy at various locations in and around the body, from the base of the spine to the top of the head; the Japanese *ki* and the Chinese *chi* refer to the life force that circulates along the pathways, called meridians, of the body (the basis for acupuncture); "orgone" was Wilhelm Reich's term for the life force. After several years, I recognized an amazing phenomenon: I discovered that I could read someone's energy pattern by moving my hands about an inch above the body. For example, "stuck" energy has a dense and stationary quality. When clients work through feelings and memories connected to a part of the body, the quality of energy changes rapidly. Previously blocked areas pulsate; muscles soften and relax. Slowing our pace and practicing with "listening hands" opens the door for experiencing this energy field, this life force, which resides in all of us.

8. Touch is a viable, accurate system of communication.

It is a sophisticated language that can both listen and communicate. It bypasses words and rational concepts and supports the growth of the nervous system, including the development of the five senses, and the abilities to orient in space, move and think. Touch is crucial to life itself, not only in infancy, but through all your years. The message and quality of touch depends on the *intention* of the toucher. Touch that evokes trust and safety can create dialogues with the client's unconscious mind. In early infancy, the baby receives and understands the nonverbal language of touch long before it understands words. With an adult, the combination of touch *and* words represents the highest form of communication.

9. The body is a metaphor.

Even our casual clichés confirm this. "He's a pain in the neck." "She makes me sick to my stomach." "I can't shoulder the burden anymore." Seeing and dialoguing with parts of the body can lead to life issues; listening touch can hear the body's story. For example, one day as my client Burt lay on the table, I noticed that one of his legs was pointed to the ceiling while the other was rotated to the right. Touching and gently moving his feet verified the difference dramatically. It was as if one leg *wanted* to stay on a straight path while the other *wanted* to go off to the side. When I broached this idea to him, he laughed. "How did you know I'm struggling with being pulled in two directions in my life right now?" I knew because his body told me.

10. The body tells the truth.

When what people say and what their bodies reveal are in conflict, it is usually the body that's accurate. Learning to listen to your body's messages is essential to self-awareness, which is why the Synergist must show the client how to understand with his mind what the body is saying. This is the underlying principle that governs my work, and I'll demonstrate example after example of it as the book goes on.

11. The body is the sanctuary of the soul.

All things embody spirit, from the smallest one-cell creatures to the complex systems of the universe. The kabbalah teaches this; so does yoga and other Eastern traditions. The natural progression of integrating body, mind and emotions in Rubenfeld Synergy may thus lead to transpersonal and spiritual growth. Often, clients deal with their "soul" issues, questioning their life values in relationships, families, communities and the world. This book will treat spiritual awareness in depth, for I've noticed that more and more today, people yearn to find aspects of themselves that transcend the material life. I believe that when people work with a method that addresses the emotional and energetic blocks in their body, their spiritual dimensions tend to unfold.

12. Pleasure needs to be supported to balance pain.

Grief, anger and pain are stored in the body; so are joy, laughter and love. Rubenfeld Synergy is devoted to bringing out the positive as well as the negative, to changing the pain-addicted client to one who can recapture emotional and spiritual pleasure. In *The Hedonistic Neuron: A Theory of Memory, Learning and Intelligence,* Henry A. Klopf claims that the nervous system learns from pleasure as well as pain. According to his theory, each time we understand new information, the brain rewards us by releasing large quantities of endorphins and other pleasure-producing neurochemicals. "Eureka!" we say, and are suffused with joy. To lead a well-balanced life, clients need to contact their joyful resources as well as their painful ones.

13. Humor can heal, lighten and enlighten.

This fundamental truth about pleasure explains why humor and laughter are so much a part of Rubenfeld Synergy. When clients get stuck in a painful or repetitive loop, using appropriate humor (*not sarcasm*) interrupts their habitual pattern. Laughter can dissolve fear, free tight holding patterns, and create pathways to creativity, insights and healing. I'll spend some time on humor in this book, partly because I love to laugh and to hear others laugh, but mostly because it is such an essential tool in Rubenfeld Synergy.

14. Reflecting clients' verbal expressions validates their experience.

When clients tell me stories, I repeat their words. When I started my practice, I repeated them to myself. Later, as I grew more self-confident, I repeated them to the client, sometimes changing a word or an inflection. Feedback from clients told me how important it was for them to know I had heard their words accurately; the repetition validated their experience. I found that when I added slight changes to their words, they were able to reflect on what they had said and to take it to a deeper level.

15. Confusion facilitates change.

"Fusion" means "union." "Con" can mean either "with" or "opposed to." Thus the word "confusing" means both a pulling apart *and* a joining, both of which are vital to the process of change. I encourage clients to feel confused, because you have to be willing to be disorganized in order to get reorganized. If you are in a dysfunctional habit pattern, it cannot be changed unless it is interrupted, and interruption means confusion. We get anxious and hate it; it's often bewildering. But we cannot experiment with new, *non*habitual behavior unless we experience the discomfort of our old ways breaking apart. We cannot change without first falling into what I have called "the fertile void."

16. Altered states of consciousness can enhance healing.

Altered states of consciousness—particularly trance states—are natural phenomena and are more common than most people believe. During a trance, clients' attention may focus acutely on certain sensory modalities and internal states of being. The Synergist may see fluttering eyes and changes in breathing patterns, skin color or tone of voice. Altered states can facilitate the client's ability to contact old physical and emotional memories and can expand the Synergist's ability to dialogue with the unconscious body/mind.

17. Integration is necessary for lasting results.

Unless clients incorporate their new insights into their daily lives, they're likely to revert to old, habitual patterns. Integration within a session can take place on many levels: words and movements, sensation and emotion, memory and images. The trick is to continue and maintain it *outside* the session, melding past and present, incorporating old feelings with new, facing pain on the road to pleasure. There is no success unless integration takes place, and this cannot happen unless the client continues to practice outside the sessions. Self-mastery develops with practice.

18. Self-care is the first step to client care.

This is a principle formulated for training Rubenfeld Synergists, a kind of warning against burnout, against blurring one's own boundaries. But it holds equally true for clients wishing to pass on what they've learned to others—indeed, for anyone who is in a relationship with others: couples, families, community members. All one needs to know are the principles and "Ahas" and all one needs to do is approach the work openly and with a generous heart. As with change, the ultimate responsibility for care lies in you. Take care of yourself—spiritually, physically, emotionally—and you will find joy in the rewards of taking care of others.

As the "Ahas" and principles became more and more defined in my thinking, one concept began to grow abundantly clear on a practical level: I could help people unlock their emotions through a safe, listening touch and guide them to recognize the causes and reasons for these emotions—some of them conscious, others unconscious—through talk. The goal of Rubenfeld Synergy, therefore, is to teach people to understand and deal with their emotional, mental and emotionally caused physical problems, using gentle, noninvasive and nonsexual touch, movement and talk simultaneously.

When we begin examining the ramifications of Rubenfeld Synergy and the rationale behind its theory, you'll be able to incorporate its lessons and its methodology into your own life. These lessons will change your relationships with lovers, friends, family and community and help you reach a place of joy and emotional and spiritual contentment.

"How simple," you'll say when you've finished reading; "I can do that." And indeed you can. When the many components of Rubenfeld Synergy flow seamlessly and elegantly from one to another, the method appears simple, even effortless. I hope it seems that way to you. Yet remind yourself, as I remind myself, of the following Zen story:

"Once again, as so many times before, the Roshi took the brush and seemingly without effort made a perfect circle. The student could not contain himself. 'Master, how do you make such perfect circles, time after time? It looks so simple when you do it.'

" 'Yes,' replied the Roshi, the barest hint of a smile on his lips. 'Fifty years of simple.' "

In the end, Rubenfeld Synergy is *not* simple. But I have done much of the synergizing for you. All you need to do is practice the exercises, both alone and with a partner, to come to an awareness of your inner self that will let you experience just how unique and valuable you are.

3.

The Energy Field: Prelude to Touch

Energy is life and life is energy, from single-cell amoebas to the most complex living organisms on earth. Indeed, though energy has been called many different things—metabolism by the biologist, kinesis by the physicist, creativity by the philosopher, fire by the poet—each is a manifestation of a single deep process. Just as we "see" life, so we can train our eyes and hands to "see" energy. As breath is visible on a mirror, so energy is palpable and heard by the listening hand.

Inside the human body, food and oxygen become energy when they are metabolized in the liver and sent to our organs so that they can function. Scientists have long been able to map this inner process of metabolism. They know which combinations of foods burn most effectively, how breath changes our physical states and why movement and exercise are essential in keeping our energy levels high.

In the early 1960s, I was invited to a meeting of body healers. They were discussing an energetic phenomenon that I had felt with my hands but did not see or understand. Apparently, ancient medical practitioners were aware of how our bodies retain, emit, flow and vibrate with energy. Indeed, these practitioners developed detailed maps of the various energy systems and how

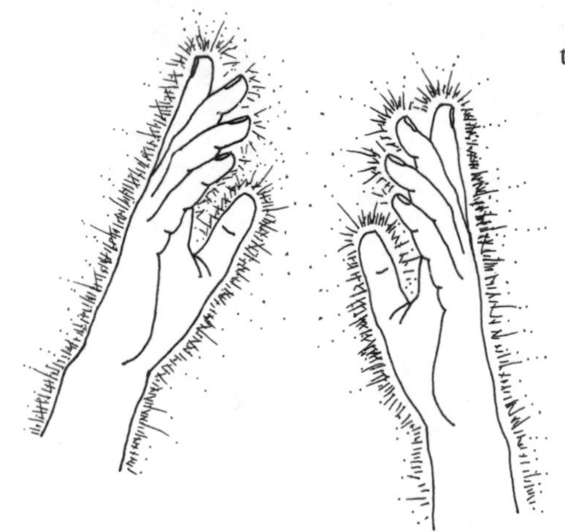

they affect our emotions and health. At this meeting, I was introduced to the concept of the energy field (or aura).

Actually I had, in a non-formalized way, experienced and listened to this energy field for many years. Now, as I practiced and repeated some simple exercises with my clients, my awakening became sharper and the energy field more dramatically noticeable. Recently, Kirlian photography has documented the energy field pulsating around the body.

Here is an exercise that will start you on your journey of listening to your own energy field:

EXERCISE: YOUR OWN ENERGY FIELD (SOLO)

Preparation: Have a notebook, pen and crayons nearby.

1. Find a quiet place, sit comfortably and close your eyes. Now vigorously rub your hands together, then shake them out gently. Float your hands up in the air with your palms facing each other, creating a space between your hands of shoulder width. Relax your shoulders and let your elbows bend.

2. Imagine that the space between your palms is the middle of a widened accordion. Slowly move your palms closer and closer until your hands almost touch, and listen to what is happening in the space between them. Now slowly move your palms apart and notice what happens. As though you were playing the accordion, move your palms closer and farther apart.

People have different ways of experiencing the energy field between and around their hands. Some see colors in their mind's eye, hear tones in their mind's ear and feel vibrations, lightness or density in their hands. What do you feel in the space between your hands? Tingling? Density? Magnetlike sensations? This time, move your hands farther apart than before. Does the wider space between them change the quality of energy exuding from your hands?

The energy field may change depending on your physical condition, your emotional state, your attitude and what you are thinking.

3. Close your eyes again, rub your hands together vigorously and shake them out. Float your hands up in the air, turn your palms in the direction of your face and move them toward it. Stop an inch or so in front of your face and move your hands up and down.

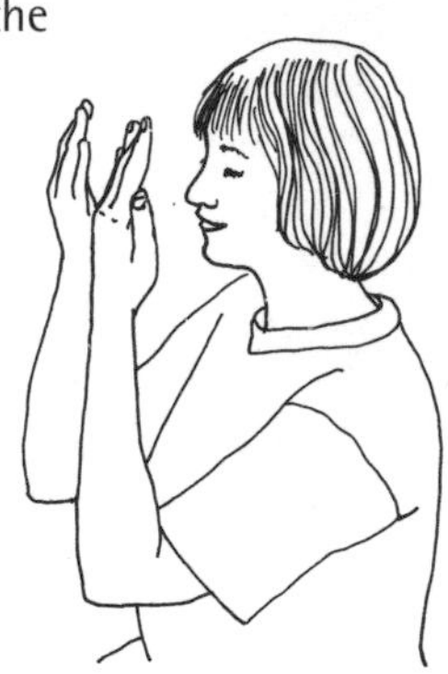

What do you experience as your hands move closer? Do you feel how the skin of your face responds to the energy emanating from your hands? Now allow your hands to move through the energy field and touch your face. Take your time and explore your forehead, then your cheeks, chin, nose, eyelids and mouth. Pause and take a deep breath. How long did you sense your hands approaching your face before they touched it? How do you experience touching your own face in this way?

Some people report that they felt as if it was the first time they'd ever really touched their face. This way of approaching seems to make a big difference and affects the quality of skin contact.

4. Pause for a moment and take a deep breath. Slowly move your hands away from your face and shift them around to the back of your head, remembering to keep a space of about an inch or so between your hands and your head. Now let your hands pass through the energy field. Touch and move all around the back of your head. Because our eyes, nose and mouth are located in front, we tend to be more sensitive to and aware of what's happening in front rather than in back of us. We usually do not pay attention to the energy field and sensations traveling toward us from behind our heads. But this area is important. Cradling the back of a baby's head is one of the earliest messages of support and safety. We will see how this touch is used in Rubenfeld Synergy.

When do you sense your hands approaching the back of your head? What do you feel when your hands touch your head? What differences do you notice between touching the front and back of your head? Take a moment to rest your hands and arms and take a few breaths.

5. Now rub your hands together again and shake them out. Float them up in the air and move them toward your ears very slowly. Stop several times on the way until you are just about to touch them. Use the accordion image and experiment with the space between your hands and ears. This time move your hands through the energy field and touch your ears. What do you feel when your hands approach your ears and the sides of your head? Are there any differences between approaching from the front, back and now the sides? Pause for a moment, rest and take a deep breath.

6. Move your hands over the top of your head. Bring them close to the crown. Now pass through the energy field and touch it. Shift your hands to several places on the top of your head. Do you sense your hands approaching? How do you feel when your hands touch your crown? Were there any differences between the approach of your hands to your face, back and sides of your head? How do you experience the rest of your body after spending this time around your head? Pause and rest your hands and arms. Take a few breaths.

In your journal, write any impressions you have about this exercise; this will help your neocortex to organize your experience. Draw any visual patterns that emerged from your hands and head.

In this exercise, you have been developing a capacity to "listen" to a non-verbal phenomenon. Practicing each repetition may bring you a different experience and awareness. Since the body, emotions and spirit are inseparable, the energy field may feel quite different depending on your mood and what you may be feeling and thinking. The important element in this first entry into Rubenfeld Synergy is that all life is surrounded by an energy field and by *listening* to it we can begin hearing the path to healing. Once you can listen to your energy field, you can listen to and hear another's. The ability to listen and hear truly opens up the possibilities of relationships with others—and a true understanding of yourself.

The Physical Energy Field

The energy field/aura is discernible in several layers. The closest around the body is the most dense and represents our physical state. The physical energy field/aura changes depending on many health patterns such as sleep, food, emotions, pain, pleasure and various degrees of stress. Sometimes the physical energy field varies in temperature and may become quite weak because it reflects an internal health and/or emotional crisis.

Since you have had some experience with listening to your own energy field, I want you to try this next exercise, which will enhance your ability to listen to another person's physical aura. You will also have a chance to check out the relationship of what you are sensing to what may be happening within your partner.

EXERCISE: SCANNING THE ENERGY FIELD (DUET)

Preparation: Ask someone to join you in this exercise. Find a quiet place and arrange not to be interrupted. For purposes of these directions, you will be the *scanner* and the other person will be the *partner.* You will be reversing roles after the first round.

1. *Partners:* Lie down on your back. Stretch your legs out. Rest your arms at your sides and close your eyes.

Scanners: Find a comfortable way to sit alongside your partner so that you will be able to move around. Rub your hands together and then gently shake them out. Close your eyes and take a few deep breaths. Ask yourself to have an attitude of curiosity and openness. There is no right or wrong in this exercise.

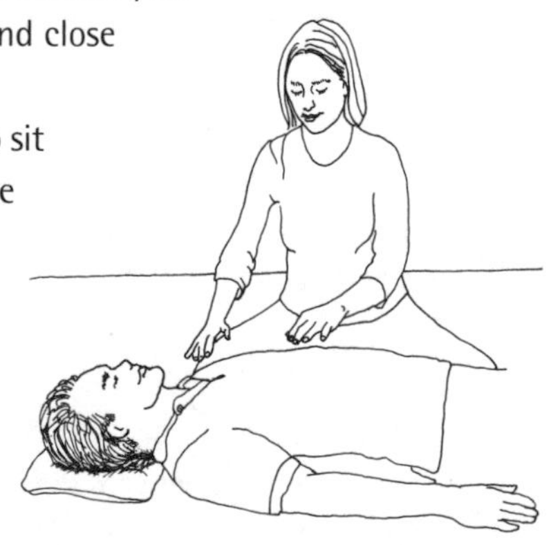

Open your eyes and move your palms an inch or so above your partner's body. Slowly move them up toward your partner's head; then around the top of it; over the face; downward toward the chest. From the chest move down each arm and over the hands; over the pelvis; and down each leg and over the feet. Then return upward from the feet all the way to the space over the head.

2. *Partners:* With your eyes still closed, imagine traveling inside your body and focus on any areas within you which may be calling out for attention.

Scanners: Continue moving your hands about an inch or so over your partner's body, noticing any sensations of heat, coolness and density. Make a mental note of these—without judgment or conclusion. Once you have completed a full energetic scan (perhaps returning a few times to several areas), slowly move your hands away from your partner, shake them out and rest them on your lap. Wait a few moments and take a deep breath. Ask your partner to open his/her eyes and make contact with you.

3. *Scanners and partners:* Once you have made contact, wait a few moments and then begin to share anything you wish about the experience.

Questions to Guide You:

Scanners:

- What was the experience like for you?
- Did you discern any differences in temperature—e.g., hot, cool, warm, etc.—while you were scanning your partner's body?
- Did you feel a change when your partner journeyed inside and focused on particular areas of his/her body?
- Did you have a sense of where your partner was focusing?
- How did you experience your hands as you moved down your partner's body? Were there any particular sensations in them and did these sensations change?

Partners:

- What was this exercise like for you?
- How did you experience hands moving over your body without touching you physically?
- When you imagined going inside, where did your journey take you?

- Which part(s) of your body called for your attention and focus?
- Did you sense the energy of your partner's hands over the area you were focusing on?

Scanner and Partner:

- Take some time to discuss your experiences and discoveries.
- Check out your perceptions of what was occurring with each other and whether or not they are congruent.
- You may want to write down your reactions and experiences in a journal and, if you wish, compare them with the other person's.
- Take two sheets of paper and draw how you imagine your body looks before and after this exercise.
- How do you both feel right now?

Now reverse roles. The partner will be the scanner, the scanner the partner. Start from the beginning.

The Emotional Energy Field

Pulsating alongside the physical aura is the emotional one. Although this layer is distinct, it moves in waves and interconnects with the physical aura constantly.

Think of storytellers. They have a special quality about them. As they spin the threads of a tale, listeners begin to identify with universal themes, empathize with the characters and feel the emotional energy field of the narrator. Listeners may cry, laugh, become angry or sad. Storytellers are performers. They know how to sense the emotional energy field of an audience through its reactions. I'm sure you've experienced a performer who says dramatic things in a dead voice. The result seems "flat." Similarly, the emotional energy field of an audience is keenly felt by the performer.

As a conductor, I quickly sensed the emotional energy field of an audience—just as I now can discern the emotional aura of an individual client, couple or group. This emotional aura is the silent partner to talking and looking. Some people have strong emotional energy fields and everyone notices them the moment they enter a room. You may also notice that others talk and act "as if" they are enthusiastic, but their energy field is minimal, low. Still others claim not to be intruding upon your "space," and before you know it they're in your face.

At the Esalen Institute, the dining room is always crowded. There are times when I don't feel like making contact or having a conversation, so I sit in a more secluded section. From the corner of my eye, I see some people approaching. They walk right over. "I hate to disturb you, BUT . . ." One of them is already in my energetic space, and I feel put upon. Whatever thoughts and emotions I may have been experiencing are shattered. The incident reminds me of an old saying: "If you let the head of a camel into the tent, the rest will soon follow."

Yet the "sharing" and "interlacing" of energy is the basis for any human relationship. By your ignoring or not sensing another's energy field, your chances of meaningful communication become slim. Love, sex, friendship, compassion, communion, laughter, joy, mutual spirituality involve energetic states. So do rage, fear, distaste, repugnance and hatred. "Why do you like him?" I ask a friend of mine about a man I find particularly unattractive. "He gives off good vibes," my friend answers. His energy has a different effect on her than it does on me, so I simply wish her well and marvel at the variety of human interaction.

Relationships are duets, with each member singing an equally important part. In a musical duet listening to one another is vital. Eventually musicians tune themselves to each other and develop an important energetic relationship. The more *we* practice in a relationship, the more profound and rewarding it will be, as our energy allies itself and harmonizes with our partner's.

Just as the body, mind, emotions and spirit are all interconnected, so are all the energy fields vibrating around the body. The energetic layers are not confined in carefully demarcated patterns; rather, they combine with each other even as they have their own space.

Some healers have actually been able to see the differences between the physical, emotional and spiritual layers of the aura. Dora Kunz, codeveloper of Therapeutic Touch, was able to see and differentiate between the various energy fields at an early age. She is quite accurate in relating the physical energy field to what is happening internally in the body. And Dr. John Pierrakos, founder of Core-energetics, used Kirlian photography to show different-colored energy fields pulsating out of leaves and flowers—and people's hands and fingertips. The colors, strengths and width of the aura changed depending on what individuals were feeling and thinking.

Sensing your own emotional energy field and its relationship to another's takes practice. The first step is to become aware of your own aura, as in the first exercise. The next is to develop your ability to listen to someone else's physical aura, as in the second. In this next exercise, I'll describe a way for you to experience and learn more about your energy field and its relationship to another's.

EXERCISE: ENERGY RELATIONSHIPS (DUET)

Preparation: Select a quiet place with ample room to move around. You will both be standing.

1. Choose a partner—a family member, friend or even an acquaintance. One of you will be called "A," the other "B." Stand about six feet apart, facing each other.

2. A, close your eyes and stand motionless. B, close your eyes and walk slowly toward A. The moment you (A) feel uneasy or sense that B is too close, say "Stop!" and take a few deep breaths.

3. A and B, open your eyes. Share whatever you wish about this experience with each other.

Questions to Guide You Both:

- When did you sense each other's energy field?
- Where did you feel the energetic contact in your body?
- A, if you became uncomfortable, did you wait to say "Stop!" or react immediately?
- B, when did you feel the energy field of your partner?
- B, how did you feel as you moved closer to A?
- B, did the energy field around A change as you got closer?
- What was the distance between you when A said "Stop!"?

After discussing these questions (or similar ones of your own), pause and move apart again before continuing. Now repeat the exercise with your eyes open. A will remain quiet while B moves. After completing Steps 1, 2 and 3 with your eyes open, share anything you wish about the experience so far. Use the questions as a guide to your feedback.

Now you are ready to continue the next phase of the exercise.

4. Stand about six feet apart. A, remain still and close your eyes. B, walk around to A's left side, then make sure you are looking directly into A's left ear.

5. B, begin moving toward A, approaching from the left side. When you (B) feel A's energy field, pause and wait.

6. When you (A) sense B approaching, acknowledge it verbally; if you remain silent, B may continue to move toward you. A, say "Stop!" if you feel B has gone beyond your comfort zone.

7. B, continue to move until A says "Stop!" or you make physical contact. A, open your eyes and see where you are in relation to B. Pause and share anything you wish about this phase of the exercise.

Questions to Guide You:

- How did you (A) experience B's approaching you from the side?
- Were there differences between this experience and B's approaching you from the front?
- When did you (B) sense your partner's energy field?
- When did either of you go beyond your level of comfort?
- If you did, how did you both notice this and where did you feel it in your bodies?
- Did you (A) allow B to go beyond your energetic boundaries before you said "Stop!"?

Take a short break and then repeat Steps 4–7 with your eyes open. When you complete this part, pause and share anything you wish about doing this exercise with open eyes.

There seems to be a distinct difference between the right and left sides of our bodies. So let's take a few moments and do the same exercise with B approaching A's right side. You may want to do this with your eyes closed and again with your eyes open. When you have finished, pause and share anything you wish about this experience.

Questions to Guide You:

- A, how did you experience B's approach from the right side?
- A, was there any difference in B's approach from the right side compared with the left?
- A, when did you sense B's energy field?
- B, when did you sense A's energy field?
- When did either of you go beyond your level of comfort?

Just as there are energetic differences between being approached from one side or the other, so there are differences when being approached from the back. Let's continue this exercise with yet another variation. Begin again standing about six feet apart. Do it once with your eyes closed and once with your eyes open.

8. A, stand quietly as B walks behind you and faces your back. Now, B, walk slowly toward A's back.

9. A, when you sense B's energy field let B know it. If B continues beyond your comfort level, say "Stop!" Pause for a moment.

10. A, turn around slowly and face your partner. If your eyes were closed, open them. Now share anything you wish.

Questions to Guide You:

- A, when did you feel B approaching your back?
- A, how close did B come to you before you sensed B's energetic field?
- Were there any differences for you both between being approached from the front, sides and back?
- B, when did you sense A's energy field?
- B, did A's energy field change when you got closer?
- How do you both feel right now?
- Do you sense each other's energy fields in a sharper, more vivid way?
- Have your self- or body images changed?
- Where are you most sensitive to another's energy field—front, sides or back?
- What sensations do you have in your body? A tingling? Density? Warmth? Coolness?

Now take a break, close your eyes and breathe deeply. We are going to reverse the roles of A and B.

In the previous instructions, B was the mover. Now B will stand quietly while A does the moving. Follow the steps in the same order. Because your roles are reversed, your process and conversations may be different. After you have completed doing the exercises in reverse roles, you may want to try a variation:

11. Stand facing each other about six feet apart with your eyes open. A and B, begin walking slowly toward each other. Continue walking until you both feel the energy field of your partner. Then either of you may say "Stop!" and notice where you are in relation to the other. One of you may have said "Stop!" before the other felt the energy field. The difference in your perceptions may bring up a variety of emotions. Take time to share them.

Here are some variations on the theme of walking toward each other:

Variation 1: Repeat Step 11 with your eyes closed. When you finish, open your eyes and process what happened. What were the differences?

Variation 2: Repeat with your arms and hands extended in front of you. Experiment with your eyes open and then closed. Upon finishing, process what happened and discuss the differences.

Variation 3: Repeat the exercise at different speeds. Notice what happens to your energy field when you move quickly toward each other. What are the differences?

Variation 4: Create a space of about six feet between you. Now stand with your backs toward each other and begin slowly walking backward until you feel your partner's energy field. What happens when you both go through the field and stand touching back to back?

As you repeat these exercises, you will become sensitized to the physical and emotional energy fields of your partner and yourself.

The Exercise in Action

These energy fields may appear elusive, but they can play a dramatic role in our lives—as an incident at a recent workshop reveals.

Eric and Emily are a couple in their fifties who have been married for twenty-five years. Emily decides to be A as Eric, B, does the moving. Eric approaches her. "Close your eyes and focus on your body," I tell Emily, "particularly any place that's crying out for attention."

Eric approaches her and stands a few inches away. He then scans her body. Following my instructions, Eric places one hand near her chest, the other near the crown of her head. Suddenly he withdraws his hands as though he had touched a live wire.

She bursts into tears.

"What's the matter?" he asks, his voice perplexed and sad.

She looks at him resentfully. "That's just like my father," she says. "You pretend to touch me, but you aren't really there."

I ask her to elaborate. She tells the group that when she was a little girl her mother went to work every day, leaving her father to stay home and care for her. But he was an alcoholic and would disappear, often for the entire day; she never knew if he would come home or not. He wasn't there for her at all, leaving her terribly lonely and frightened. When Eric pulled his hands away without warning, these feelings flooded back into her body.

She turns to him again. *"Why did you take your hands away?"*

"I had a different kind of father," he whispers, his eyes filling with tears. "Almost as bad. He was cold, cold. He never touched me."

Emily's gaze is now sympathetic. Touching her back, I ask her to look into Eric's eyes. "Who is he?" I ask.

"Eric, you're *not* my father," she says softly. "I want you—I need you—to stay with me. Don't leave so quickly."

"Can you stay with her?" I ask him.

"Yes," he says, but he is visibly anxious.

Now I am touching his back, which is braced and tight. I press him further. "Can you say, 'I'm not going to be like your father'?"

He returns his hands gently near her chest and head. "I'm not going to be your father," he says. "I'm Eric." His back loosens, relaxes.

She closes her eyes and sighs with pleasure. He keeps his hands near her for about a minute, until she moves away. When she opens her eyes she looks directly at him and smiles.

The group around them, feeling Eric and Emily's emotional aura, are sobbing. They have felt the enormous waves of energy coming from both of them—and now burst into applause.

I place my hands at the base of her spine and on her belly, feeling the vibrations of her emotions. "I think we've turned a corner in our marriage," she tells me. "After all these years—now he can empathize."

Note that the exercise Emily and Eric did involved scanning, moving through the energy field and then touching. Eric's cautious approach and his contact with Emily's emotional aura led them to a deeper understanding about the dynamics of their relationship.

The next exercise I want you to do will also develop your intuition and open another gateway to your inner journey. It is the first one in which scanning leads to touch. You will simply allow yourself to be present, without a preconceived agenda; there is no right or wrong way to do it. There are no mistakes, only discovery. You will be using a part of your brain that doesn't have to know "why" at this moment. If you keep an open mind, without judging yourself, your hands will sense what is happening and they will lead you. My only caveat is that you both agree not to touch the other's breasts or genital area, and that if your partner says "Stop!" you will immediately take your hands away.

EXERCISE: SCANNING AND TOUCHING THE EMOTIONAL AURA (DUET)

You may want to do this exercise standing, sitting or lying down. In the following instructions, you will both be standing.

Begin with you as A and your partner B. Find a quiet place and take your time.

1. A, stand with your eyes closed and your arms hanging loosely at your sides. B, stand a few feet away, facing your partner. Both of you remain still for a few moments and take a few deep breaths.

2. B, begin to move toward A until you both find a distance between you that is comfortable.

3. B, rub your hands together and shake them out. Begin the scanning process (Exercise III), remembering to keep your hands about an inch or so from A's body.

4. A, let your memory go to a scene that evokes emotions. See that scene in your mind's eye. After a few moments, notice where the emotions reside in your body. Continue to silently focus on this area.

5. B, continue to scan A's body. As you move your hands, notice any difference from one place to another. Without asking "why," pause and let your hands lead you through A's energy field and gently touch that area.

6. A, experience the gentle touch of your partner and allow any emotions to emerge. Take a deep breath.

7. B, continue to touch for a few more moments and when you feel ready slowly move your hands away, shake them out and let them rest at your sides.

8. B, make sure you are facing A so that you are in full view when A opens his/her eyes.

9. B, greet A, your partner, back from the journey.

10. A, slowly open your eyes and let yourself see your partner, who is welcoming you back. Share the memory and scene if you wish. Allow enough time for both of you to share whatever else you might want to.

Questions to Guide A:

- When did you sense your partner approaching?
- What energy did you feel when B scanned around your body?
- Did you feel any emotions as you remembered your scene? What were they?
- Where were the emotions located in your body?
- How did you experience the touch of your partner?
- Did this touch affect your emotional state? If so, how?

- How did you feel when your partner moved his/her hands away?
- What discoveries you made about yourself are meaningful and useful?
- Can you use these discoveries in your life?
- What are you experiencing now as you complete the exercise?

Questions to Guide B:

- What was the distance between you that felt comfortable?
- When did you feel ready to scan your partner's energy field?
- Did you feel any differences as you moved your hands closer to your partner's body?
- When you passed through the energy field, where did your hands guide you to touch?
- Did you experience any images or emotions when you stayed in one area for a while?
- What did your partner look like when you took your hands away and faced him/her?
- What are you experiencing now as you complete the exercise?

Take a break and rest. This is a potent time to write your reactions, feelings and ideas in your journal. You may want to express yourself by drawing.

When you complete writing and/or drawing, switch roles. B, stand quietly, close your eyes and go inside your body. A, stand a few feet away, approach your partner and follow the directions of the exercise on page 36.

Intuition plays an important part in all of these exercises, but especially in this one, because you are listening to each other without words. As the scanner, you do not receive information in the usual way, and therefore your ability to listen to another person is increased and deepened. Since body, thoughts and emotions are one, touch and its energy serve to connect any physical and emotional areas. People may experience energy within themselves that is far from the source of contact. Indeed, listening hands may begin a flow of energy to distant places that are calling out for it.

Ancient and modern healers have learned how important energy flow is and they recognize that it is not random. There is an energy system within the body that is governed by the chakras.

The Chakras: The Seven Steps to Aliveness and Spirituality

For many years, though I sensed distinct energy clusters throughout the torso, neck and head, I didn't refer to them as chakras. Presenting Rubenfeld Synergy to therapists and allied medical practitioners was enough of a challenge, and I didn't want them to view my method as "strange" or "Eastern." It was actress Shirley MacLaine's discussion of "auras" and the "seven chakras" on *The Tonight Show* that popularized the chakra system and introduced millions of Americans to these ancient ideas. I had to smile watching her pin seven different-colored circles onto the back of Johnny Carson's clothing. Today, most healers talk freely and without embarrassment about energy spiraling up and down the spine.

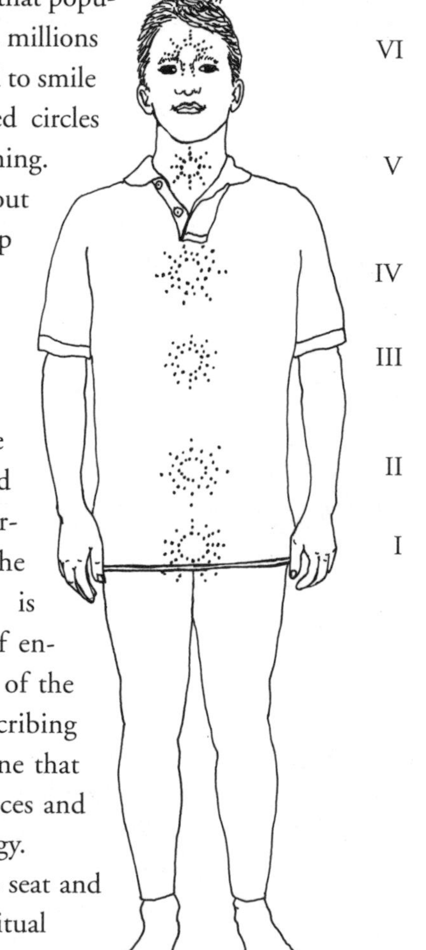

The notation and mapping of energy radiating from within the body was documented over ten thousand years ago by Vedic scholars and healers. The energy field radiates outside and around the body, but its source begins inside, organizing itself in specific locations. The Sanskrit name for this phenomenon is "chakras," meaning "spinning plates" of energy. There are actually several versions of the chakra system, and many books describing them. In this chapter, I'll refer to the one that had the greatest impact on my experiences and on the development of Rubenfeld Synergy.

The seven chakras are considered the seat and source of physical, emotional and spiritual energy.

CHAKRA I: The source of this energy begins its journey at the base of the spine—called the root or first chakra. When the first chakra is activated it governs survival (flight-or-fight) instincts and procreation.

CHAKRA II: The second chakra is located in the pelvis, an inch or so below the navel in the front, and in the lower back. It governs sexual energy, power and creativity. In many Eastern martial arts, this area is called the *chi,* as in tai chi, or *ki,* as in aikido, and is regarded with great respect because it is the source of power and strength.

CHAKRA III: This chakra is located in the solar plexus (the stomach area in the front) and in the middle back. It governs the emotions, which greatly affect how the intestines absorb and digest nourishment.

CHAKRA IV: This is known as the "heart chakra," the place of universal love—of family, animals, friends. Our expression "Open your heart" is in direct lineage from the fourth chakra. I'll be discussing what happens to its energy when the chest is tight and breath is shallow.

CHAKRA V: The fifth chakra lies in the throat and manifests itself through the voice and its quality. When the throat and neck are tight, sound is difficult to produce and people have trouble in the vocal expression of what they are thinking and feeling.

CHAKRA VI: Found between the eyes and known as the "third eye," it carries ancient wisdom and knowledge of phenomena beyond the visible and obvious. Sometimes events and information do not make rational sense, yet the third eye sees and understands them.

CHAKRA VII: This chakra is what all spiritual beings aspire toward. It is the energy vibrating in and above the crown of the head. Unconditional love and regard for all humankind, earth and the cosmos emanates from this place. In Eastern religious art, it is represented as a fully open lotus flower; in the West, by a golden halo.

When the chakras spin, they create energy, movement and a great deal of heat. Scientists have witnessed Tibetan monks meditating in temperatures of 50 degrees below zero wearing only light sheets. When examined, the monks were found to be sweating and extremely hot. The monks explained that they knew how to cause the chakras to spin with great force and move heat throughout their bodies.

In a healthy body, energy begins from the root chakra and flows upward through the crown chakra. Imbalance occurs when the energy flow is blocked; Eastern healers believe that this energetic imbalance is the cause of disease. From this point of view, a stomachache or digestive problem, for instance, may result from blocked energy in the third chakra. Since this is also the chakra for emotions, it may be stress and unhappy life issues that are impeding the energy flow. Teaching people how to listen and recognize the sensations of these blockages is the first step toward melting them and allowing the energy to flow through. Indeed, this is one of the goals of Rubenfeld Synergy.

As you practice and repeat these exercises, you will notice that energy patterns change. Pulsations, texture and density of energy alter depending on the mental and emotional states of a person. A calm, meditative and peaceful state of being produces an energy pattern which is distinct from that of someone who is experiencing fear or anger, joy or sadness.

Many people across the world live in the first chakra. Their life is primarily about survival—feeding their families and procreating. Others live in the confines of the first and second chakras, from survival and procreation to power.

There are those who do not have material wealth, but are rich spiritually. Many shamans (traditional healers) have the extraordinary ability to experience and live in energetically altered states, gaining wisdom in a way that may not be obvious or understandable to us. The movie *Star Wars* spread the concept of using focused energy, not brute force, to win and heal. In my practice, my hands are able to feel the links between these emotional, physical and spiritual energies. This sustains my conviction that the act of contacting, experiencing and releasing thoughts, emotions and body tensions allows life's energy to reach a realm full of love.

Acupuncture: The Internal Maps of Energy

At first, Western medicine would not acknowledge the concept of "invisible" lines of energy. Now it has finally accepted the evidence that acupuncture works as a complementary form of diagnosing and healing. Acupuncture has been successfully practiced in the East for thousands of years. The ancient Vedic doctors mapped the energetic routes as large rivers with smaller streams throughout our bodies. Maintaining energetic balances—hot and cold, light and dark, masculine and feminine—is the acupuncturist's concern. It is not my purpose to explain acupuncture in detail here (there are many books available), but rather to relate it to the "listening hand."

For example, the fourth chakra is located approximately in the upper chest—around the area of our anatomical heart—and between the shoulder blades. When I slide my palms underneath one of the shoulders, I can feel tensions and holding throughout the torso. As the client and I talk, my hands are also listening. They discern physical and energetic changes. With time, there is a "letting go" of tensions and an expansion and opening around the heart. (It's no accident that the Eastern scholars' name for this meridian, or energy line, that my hands are touching is "the gateway to the heart.") A master acupuncturist who trained in Rubenfeld Synergy showed that the body's energetic balance could be reestablished without using needles, but by instead using touch.

You can see some children with rounded backs and collapsed shoulders constricting their heart space. If we paid attention to such physical metaphors early on, we might be able to help these little ones change the course of their emotional lives.

One major goal of this book is to teach you how to listen to early stages of imbalance and the first signs of pain. But what is pain? Is it physical? Mental? Emotional? Spiritual? There have been anatomical, psychic and psychological explanations. Whatever the answer, pain is an early and truthful warning signal that something is going on which needs attention. In the field of psychosomatic medicine, physical, mental and emotional pain are interconnected. They *all* need to be considered if we are to heal the whole person. This possibility is greatly enhanced by the use of touch.

4.

Listening Touch®

I had never witnessed the birth of a baby until my godson Adam was born. His parents, Bill and Margaret, were Rubenfeld Synergists and this was our first "synergy" baby. Honored and excited, I arrived at New York's Mount Sinai Hospital and was ushered into a wing specifically planned for parents who wanted a midwife to guide their baby's birth in a homelike environment. For me, touching Margaret was a delight. Her energy field was so bright and intense I was able to capture it in photographs. The atmosphere was calm, the lights muted, the walls painted a soothing pastel blue and the bed particularly comfortable and commodious. Music from Mozart's *Magic Flute* pervaded the room.

For months, Bill and Margaret had sung and talked softly to the baby inside Margaret's womb. Now we all took turns talking to him, welcoming Adam into the world and his family. He could not understand the words, but the vibration of our voices, along with the warm, soft flow of amniotic fluid, were the first gentle and loving touches he received. If an unborn baby can "feel" safe, that was Adam's condition.

Margaret chose natural childbirth and took no drugs. She labored mightily. My heart began to beat rapidly as the top of Adam's head appeared. Urged on by our words of encouragement, Margaret pushed harder. Finally, Adam's head came sliding out, and the rest of his body followed. Cradled by the midwife, he opened his eyes and cooed.

The midwife placed Adam on his mother's chest. Margaret held him lovingly, murmuring sounds into his ears. His umbilical cord was still pulsating. The midwife gave Bill a knife and he cut it.

He removed his shirt and lay down beside Margaret, moving the baby slowly onto his chest so his son could hear the thumping of his heartbeat. Then it was my turn to hold the baby. Though I had never borne children myself, every part of me recognized the miracle of life's creation.

Adam simply gazed into our faces and continued to coo and make other sounds of contentment. Mozart's music played on, filling the room with harmony.

Under New York law, each newborn must be examined by a physician and drops of silver nitrate must be placed in his eyes to prevent infection. The midwife immediately called Margaret's obstetrician, who happened to be jogging in Central Park near the hospital. He arrived a few minutes later, still dressed in his running clothes. Smiling, friendly, obviously pleased the birth had gone so well, he touched the baby.

Adam howled!

For all his goodwill, for all his friendliness and warmth, the doctor's *intention* when he touched the baby was to examine him. He was doing his job and doing it well, moving Adam this way and that, peering into his ears and shining a light into his eyes. His relationship to Adam was of scientist to subject, doctor to patient.

Adam felt and knew this instantly. His contentment turned to rage, and when the drops went into his eyes, to outrage. Until the moment the doctor held him, Adam had experienced only the touch of loving, caring, emotionally committed hands. That there was another kind of touch—matter-of-fact, peremptory, neutral—aroused such howls of disenchantment that it seemed to take forever to calm him.

Newborn infants do not understand words; they hear the vocal tone and music of the voice. But they *do* understand, experience and respond to the toucher's message and intention.

Watching Adam's reactions, I remembered my first sessions with Judy Leibowitz. My struggle to understand her words left me feeling frustrated. Eventually, I realized that her special touch contained the message—not her words.

My body began to understand not because of what she said, but because how she touched those tight, frozen places within me.

In the 1950s, Dr. René Spitz, working with orphanage babies, discovered that the holding, cuddling and touching of newborns and growing children is literally a matter of life and death. The orphanage babies died or suffered severe mental defects because they were not lovingly handled. Their nervous systems were not stimulated to grow. Their skin did not come into contact with human warmth.

Adam would have no such problem.

The Meaning of Touch

Eskimos have many words for snow. Since snow is an essential in Eskimo life, it must be referred to with precision to connote its significance to the speaker and to the listener. There is snow for hunting, snow for traveling, snow for building, snow for melting and drinking, etc. Touch doesn't seem to have an equivalent importance in our society, for we have only the single word to describe myriad kinds of touch, myriad significances.

Although there are as many ways we use touch as the Eskimos use snow, we are forced to rely on adjectives and adverbs. "Compassionate touch"; "sensual touch"; "abusive touch"; "friendly touch"; "she touched him lovingly"; "he touched her with care"; etc. If in musical terms touch is the theme, then the adjectives and adverbs describe its variations.

To the list I've added my own adjective: "listening."

Using a "listening" touch to hear the body's story is a powerful communicative tool. To really listen to another human being and yourself, not just in words, proclaims a relationship that demands being in the moment, fully focused on what you are hearing.

In the past forty years, I've worked with over fifty thousand people, in private practice, groups and workshops. A major theme plays its refrain repeatedly: *People from all backgrounds yearn to be listened to.* When people contact their early childhood memories, they usually grieve for those times their parents didn't listen, didn't *hear.*

Touching another person and yourself with clear intentions is basic to the philosophy and practice of Rubenfeld Synergy. Touching is by itself neither good nor bad. The message of the touch depends on the intent and purpose of the toucher and receiver and the context of the situation. It's a two-way channel, a circuit, a loop of communication—a duet—through which your hands listen, receive and "speak" to another and yourself. In fact, touch and healing communication have existed from the beginning of humankind. For thousands of years, the "laying on of hands" for healing was a custom in tribal life and in many religious practices. Women were considered powerful healers until the Middle Ages, when they were condemned as witches for precisely the practices that made people well. Eventually, "healing" and medical practice began to separate into different domains. Scientific diagnoses and the curing of symptoms became more important than the patients' feelings and emotions.

While traveling in the United Kingdom last summer, I was saddened to read about a law that had just been passed, making it a crime for any teacher to touch any child, from prekindergarten on. In the name of protecting our children from abuse, I thought, we are depriving them of life-giving contact. The article went on to describe the frustrations of many teachers whose natural inclination was to hold a crying or upset child, but were hesitating to do so because of their fear of prosecution. Teachers had even asked whether they should apply sunscreen to children when they were patently in danger from the sun. This law is surely a sign that we are becoming further removed from the natural nourishment and connectedness of touch.

Technology has helped people to reach each other without being in one another's presence. Relationships are conducted in front of computer screens, and E-mail allows them to "talk" without experiencing the vibrations of each other's voice. Although I'm not against this technology (its advantages are obvious in certain contexts), I'm alarmed by the speed with which people have divorced themselves from a primary way of making contact with other human beings.

The listening hand brings us back to an ancient kind of language. This kind of knowing, contacting, understanding and hearing adds depth, wisdom, compassion and healing to our lives and the lives of those we love.

Intentionality

I believe that all human beings have the innate capacity to heal. This power comes from a profound and clear *will* to heal that encompasses the mind, body, emotions and spirit. Many of my colleagues have been working with women who have cancer or are recovering from it. They have noted the marked difference in the quality of the healing process between women who are wholeheartedly committed to and believe in their process of getting well and those who lose heart, become depressed and give up. Many support groups have been created in which the participants share their stories and simply hold each other. This simple intervention has now been scientifically proven to prolong the life of breast cancer patients.

The intention to heal ourselves and others demands an ability to focus totally and to listen without interpreting or being distracted. Easier said than done. A touch with healing intentionality comes from a heartful and compassionate state of mind, a willingness to be open to whatever presents itself. If there are any thoughts of "Let's get this over with" or "I know what should happen" or "You're not doing what you're supposed to," the person receiving your touch will "hear" these thoughts, both consciously and unconsciously.

Dr. Manfred Klynes has devoted his life's work to researching how emotions and thought messages transmit themselves through the fingertips and hands of musicians. He now has shown this phenomenon at work in nonmusicians as well. His work supports my theory that the toucher's intention and nonverbal message ("I care for you"; "I won't hurt you"; "I'm here and present for you"; "I support your process"; etc.), if transmitted clearly, affects the receiver's state of being. There is without question a powerful correlation between what the toucher is thinking and the quality of touch.

"Imagine touching someone from a compassionate place within you. Where would that place be?" I've asked this question of many workshop participants, and invariably they point to their heart. The source of this healing intention is clear to us all.

Listening from Your Heart

It takes practice to develop an ability to listen nonverbally and to transmit your intention through listening hands. As a former musician, I certainly know about practicing with patience. In music, as in Rubenfeld Synergy, the result is worth it.

Let's begin with yourself first:

EXERCISE: LISTENING TO YOURSELF WITH TOUCH (SOLO)

1. Find a quiet and comfortable place. You may sit, lie down or stand. Have paper, pens and colored markers available for use after you complete the exercise. Take a few deep breaths.

2. Imagine yourself beginning a journey inside your body. Travel around and notice which areas come to your attention. By slowing down and listening to your body, you will begin to hear a particular place calling out: "Please pay attention to me now."

- Perhaps a long-forgotten or recent memory will emerge.
- Perhaps an emotion becomes clear or more intense.
- Perhaps you will meet someone you've been unconsciously holding on to.

Remain at the area that is calling out. Take your time.

3. Now gently place your hands on that area; if there are two areas, use each hand. If that place is not easy to reach or you would prefer not touching it, *imagine* that you are.

Focus on this area or areas, allowing the energy in your hands to flow freely. In this moment, be very clear that your intention is to hear what your body is saying.

4. Close your eyes and begin a verbal dialogue with this area by asking, "What do you want to reveal to me?"

Imagine that the area has a voice and allow it to answer you. It may say:

"I've been hurting so much."

"I'm sad."

"I'm scared."

"I'm angry."

"I'm joyful."

"I'm holding a memory of . . ."

Speak the answer or answers out loud so that you can hear the responses vocally as well as mentally.

5. Repeat the answer, add the word "because," and complete the sentence without censoring yourself.

"I'm hurting *because . . .*"

"I'm sad *because . . .*"

"I'm scared *because . . .*"

"I'm joyful *because . . .*"

"I'm holding a memory of *. . . because . . .*"

How do you feel as you complete these sentences? Pause for a moment and allow yourself to experience any emotions which may bubble up.

Memories, stories and emotions are often locked in your body behind walls of denial, explanation and belief erected over the years. Practicing this specific listening touch takes down those walls and opens you to many insights and wisdom.

Perhaps your body has informed you about a situation or memory you have ignored. Perhaps it has issued a warning and demands action. Perhaps it has made a plea for appreciation. How can you use your body's message in your life?

6. Continue to listen to that area with your hands and notice if there are changes happening in your body. When you sense your body shift, slowly move your hands away from it. Take a moment to acknowledge that you and your body have listened to each other. Open your eyes slowly.

7. Take a deep breath. Write anything you wish in your journal, letting your pen flow over the paper. Sketch anything that expresses your feelings. You may even want to combine your writing and drawing.

The Exercise in Action

At a recent workshop, I asked several participants to share what had happened as they practiced the exercise. Three women from different generations responded eagerly.

Two of the women had touched their abdomens and contacted their dead husbands. "He is at peace," one said. "He tells me that I can be at peace too."

"My husband misses me," the second said. "He wants to be with me." She paused for a moment, then smiled. "He *is* with me," she announced. Then, softly, "And I am with him."

The third touched her arm. "It's tight," she said. "Painful." She held her breath and clenched her jaw.

"Speak to it," I advised. "What's your arm telling you?"

"It says I'm putting too much stress on myself and that's why it hurts."

"Is that true?" I asked.

"I don't know." She touched her arm again and winced. "I don't *think* I'm stressed."

"Tell me what you did last week."

"Oh, last week I took my law school finals. And my mother was sick and I had to visit her in Florida. And—" She began to laugh. "Maybe my arm *is* telling me something."

"Listen to it," I urged, smiling. "Maybe your arm is warning you that if you continue at this pace it will keep on hurting. Pay attention to this wisdom before the situation becomes more serious. Maybe it's time to slow down and have some fun."

The three women told the rest of us that they had never heard of the concept of listening touch, let alone experienced it. Yet all three were able to quickly learn how to utilize it and discover places within themselves that wanted to be heard.

EXERCISE: HAND/ENERGY CONVERSATION (DUET)

The usual way people communicate when they meet is by talking and looking. They gather information about each other using words and their eyes. As the encounter progresses, other phenomena occur which are not so obvious, yet are very present, such as energy patterns, vocal tone, body movement or stillness. The brain registers all of these impressions. Nearly instantaneously, but often out of our awareness, the limbic system registers emotions, body sensations and energy chemistry. The neocortex responds by organizing words, concepts and thoughts.

Since we are so used to contacting people this way, I'm going to take you through an exercise that will establish the importance of intentionality and show you an alternative way to communicate. You may invite a familiar person in your life or someone you've just met to be your partner. But remember, this is an exercise designed for your own self-discovery and wisdom. It's *your* experience that matters.

1. Find a quiet place where you can sit comfortably facing your partner, who may be a friend, lover or someone unknown to you until now.

2. Introduce yourself and share anything you wish for a few minutes.

3. Now stop talking and close your eyes. Eliminating words and vision will give you the opportunity to heighten your awareness of other ways of making contact.

4. Rub your hands together and shake them out. Then place the backs of your hands near your chest and position your palms to face your partner's palms.

5. Slowly move your palms toward your partner's. Notice what happens to the energy field between you as you approach your partner. You are moving through space, approaching another human being. Do you sense that person approaching you? There is no rush. Listen to the space between you and your partner.

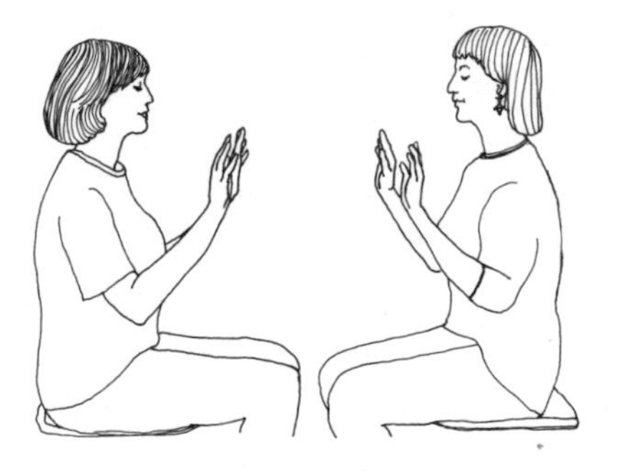

6. Allow your palms to touch. Take a deep breath and stay still for a few moments, noticing how you feel.

EXERCISE: TOUCH/MOVEMENT CONVERSATION (DUET)

We will explore several intentionalities in this touch/movement conversation:

7. Go to a time in your childhood when you played games, laughed and had fun. How might your hands express this fun, joy and playfulness as they touch and move with your partner's? Giggles and sounds are welcomed.

8. Now come to a quiet moment. With your hands still touching your partner's, breathe deeply and notice what you are experiencing in energy and your emotions.

9. Remember a time when you were openly curious. How might your hands express curiosity as they touch and move with your partner's? Listen to your partner's touch. Allow some time.

So far, you've explored two intentionalities that affect the messages that you send through your hands. The next one is at the core of all human healing and communication through listening hands.

10. Allow the words "compassion," "care" and "concern" to fill your being. Notice where you experience these words in your body. Now begin to touch your partner's hands with a clear intention of compassion, care and concern. How do you feel sending this message—and receiving it? You are sharing a compassionate, caring touch with someone whose history and life stories you may not know. And it really doesn't matter whether you know the person or not. The comfort you are transmitting at this moment reaches depths of wisdom and knowledge about another person that would have taken a great deal more time to achieve had you used only words. What the two of you are sharing is a hand/touch conversation with clear intentions. Your brain and every cell in your body may experience emotions, resistances, surprises and understanding as you practice this unspoken dialogue.

11. I invite you to continue this "conversation" for a few more minutes, expressing any feelings and thoughts that arise without words.

12. Find a way to say good-bye to your partner without speaking. Move your hands away from your partner's and return to your own space and rest your hands. Take a few deep breaths. Listen to your body, thoughts and emotions with your eyes still closed.

13. Now slowly open your eyes, allowing them to be soft and not straining to see anything. Allow the vision of your partner to come to you.

This first moment is precious. With "soft eyes," see your partner again.

Questions to Guide You:

- What do you notice about your partner now?
- Are your impressions different from what they were when you met each other at the beginning of the exercise?
- How do you feel right now?
- Are there outstanding moments that you remember about the exercise?

Your partner may look different (and you may look changed as well).

- What are some of these changes?
- What have you discovered about yourself and your partner?
- What have you learned about yourself when you are in a relationship?

Verbalize whatever you wish to share with your partner.

The Exercise in Action

Although I've led this exercise for over thirty-five years, the reactions of people still surprise and inspire me. I've used it with couples, parents and their children, strangers and pairs who do not speak the same language.

For example, in 1983 I was at a psychology conference in Moscow and was asked to lead an exercise. I chose the touch/movement "conversation." I asked each Russian to pair up with either an American or a Canadian. As people arranged their chairs to face one another, their body language was clear enough: arms crossed over the chest, one leg placed over the other, tense faces.

A Russian TV star translated my instructions and jokes brilliantly. The energy in the room began to change. As the exercise commenced, I could see they were touching and moving with each other's hands, just as you did. By the end, faces were smiling, eyes were shining, emotions were shared. People had broken through more than a language barrier—they had contacted each other from a heartfelt, playful and compassionate place. Words were not necessary. Touching with intention reached a level of safety that allowed everyone to relax and be more spontaneous. Then a bottle of vodka appeared.

The success of Rubenfeld Synergy is dependent on the ability of its practitioners to use their listening hands when they make contact with their clients. The exercises in this chapter were designed to give you an experience of listening touch and to practice how intentionality changes the quality of that touch. As you do them, notice how much attention the body requires and what it "says" in response to touch.

5.

The First Touch

When we first touched Baby Adam, it provided him with an important sensory awakening. Since he was so vulnerable and open, his responses were immediate; he had not yet developed ways to defend himself. Luckily, he experienced compassion and care.

When Rubenfeld Synergists use "first touch," people may experience a gamut of emotions. "At last, someone is touching me with respect and care," Martha says at a workshop. Fred shies away and only reluctantly lets me put my palms under his shoulders; he does not have to announce that he is afraid of contact. People may also experience waves of grief because they never had loving touch from their parents. They grieve for the lost love and now, as adults, yearn for compassionate, empathetic touch.

Taking Baby Steps

I use touch lightly, sparingly and slowly. It gives people time to deal with their feelings, express them and integrate any discoveries into their consciousness.

A first touch that is too fast and too deep may be overwhelming, re-creating the defensive walls which guarded us against feelings in the first place. One simple touch starts a ripple effect in the entire body. We are not just touching a head, arm or leg, we are touching and making contact with the whole person—body, mind and soul. Every cell recognizes this moment; it becomes the first step in healing.

The envelope of skin covering your entire body is full of sensitive nerve endings, especially at your fingertips. Indeed, fingertips receive and send messages continually. This direct and immediate contact, fingertip to skin, establishes a relationship with the unconscious mind.

Loving touch, or the lack of it, affects our lives at the deepest level. Hiding under the surface of our defensive shields may be very early nonverbal emotions of fear, sadness, joy, anger and loneliness. Touch often ignites old feelings of despair, rage or pain, but it also revives love, joy and hope. In any meaningful relationship, we depend on touching and being touched by others.

When we develop the ability to have a loving, compassionate and caring touch for others, we develop the same skill for ourselves, providing all of us with the strength and spiritual awakening that is necessary to live life fully.

Less Is More: The Butterfly Touch

The lighter the touch, the more you will hear changes happening in others and yourself. This approach is opposite to our culture's marketing approach, which shouts, "More is better!"

How much effort and strength does a butterfly need to balance itself on a flower leaf ? It uses just enough power to contact the landing place, respond to the environment and procure food. Our culture's overblown and needless use of force is like a brick dropping on the butterfly *and* the leaf, destroying the essence of the plant and the butterfly's source of sustenance.

I ask people to image a butterfly, its quality and its lightness, before touching for the first time. If your initial touch is heavy and pushy, *you* are the one controlling the relationship and telling the body/mind what to do. Instead of listening, you are already filled with the "shoulds" of what is supposed to happen (and the results) and you have lost contact with your own and another's inner feelings.

Beginner's Mind and an Empty Head

"Empty your head of any agenda."

"Let the outcome unfold."

"Be willing to experiment and remain curious."

"Call on your own life experiences."

"Let intuition, human wisdom and knowledge guide you."

These are my instructions to myself before I embark on any therapeutic process—before I start with the first touch.

Jonathan, thirty-two, rushes to the workshop table. He's eager to experience the "first touch" I've been explaining to the group and to be my subject. His black hair is tousled and his eyes shine. He's wearing white socks, a bright red shirt and a pair of corduroy slacks; his boyishness makes me smile. I invite him to lie down and he does so, looking confident and well put together, belying the fear and anxiety I am about to discover in him through my first touch. (Clients may sit, stand, move or lie down during a Rubenfeld Synergy session. However, when clients lie on their backs, it allows the tight holding areas and their chakras to surface more easily.) Before I touch Jonathan (or anyone), we come to a fundamental agreement: that he can ask me to stop touching him for any reason during the session. Then we begin.

With a beginner's mind, I make a commitment to be totally present, to detach from the outcome and to accept Jonathan where he is right now. While he settles down, I go on an inner journey of self-care, which is crucial when touching others, both at this moment and throughout the session. (I will elaborate on the importance of self-care in Chapter 10.) Finally, I am ready for the first touch.

This has always been a special moment; the first touch establishes a contact that can only grow. My first gentle touch is on the sides of Jonathan's head. Jonathan and I connect nonverbally, opening the doors to metaphors,

impressions, feelings and sensations. We are tuning into each other as instruments tune up at the beginning of a concert—sharing our energy fields.

Remember how in the exercises in Chapter 4, intentionality affected the quality and message of your touch? My intentionality is to make healing contact. Almost immediately I begin to feel pulsations rhythmically beating in Jonathan's skull, all the while watching his breathing patterns. With my fingertips, I add small movements to the touch, moving his head gently from side to side, informing myself about his state. Does he move easily or is he holding on tightly? In fact, I had noticed that Jonathan was quite flexible during a previous body/mind exercise, but now I can hardly move his head and his neck feels like tight bands of steel.

The first chakra (survival) and its message (fight or flight) is located far from my touch, but it is obviously sending out an alarm now. It's common and normal for people to have reactions of fear at the start of a session; there are many unknowns about our relationship and about what will happen. Creating a space of safety and trust is an essential component to the first touch and all subsequent touch so that clients will be able to express their feelings, reveal unspoken issues and try different behaviors.

"What are you experiencing now?" I ask.

"A feeling of being nurtured and a tingling in my arms. I started to hold my breath, like when I was a kid, but when your hands touched my head I began to relax and feel supported."

His need to be nurtured is strong and clear. If my touch had pushed him beyond his present limits, he might not have felt supported or safe. The sense of trust between us, I know, is key to our relationship.

I watch Jonathan's breathing pattern closely as I move his head. With my touch, I am sending him a clear message: "I'm here listening to you. It is not my intention to manipulate or hurt you."

From his head I move around to his feet, those expressive extremities that reveal so much about people. Feet are extensions of the first chakra and they connect us all to the earth. Ah—feet: they hold so many stories! Generally, they tell us how people "take a stand" in their lives and whether they are able to receive nourishment from their environment.

Does Jonathan stand and walk a lot? Is he a frequent runner? Does he walk pitching forward or on his tiptoes? ("Tiptoeing around" and "walking on eggshells" are two idioms that express the way people live their lives.) Your listening hands will begin to hear these stories when you pay attention to the

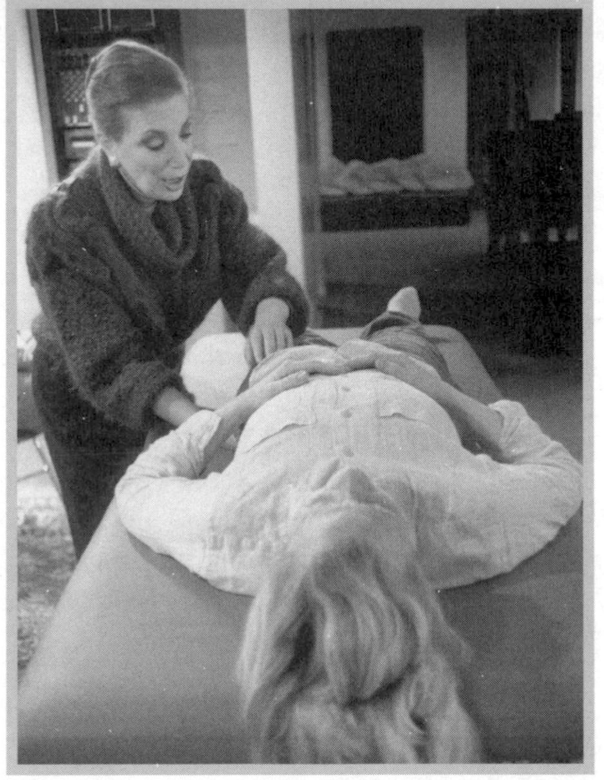

texture and quality of the feet you are touching. Are they callused? Soft? Tense? Relaxed? The answers will tell you a good deal about the client. Jonathan, I quickly discover, walks a good deal and is on his feet a lot, but is nevertheless timid, unused to striding with confidence.

"So, Jonathan, what are you aware of as I touch your feet?"

"My feet are very cold. Your hands are warm."

Cold feet! Exactly! When we explore this metaphor, it develops into a picture of a businessman who accomplishes a great deal despite his fear and anxiety, expressed by his tight neck and cold feet. He tells me that he expends so much energy to override his anxieties and keep going that he is often exhausted at the end of the day.

Now the challenge is to help Jonathan discover how to release the energy locked in his feet, move it up through his legs and connect it with his pelvis power. Gently moving his feet again, I can see how disconnected he is from them. Soon I discover that the energy is locked in his ankles; its movement stops there and goes no farther.

Learning about the state of his pelvis and hip joints through touching and moving his feet and ankles is less threatening than going directly to his pelvis early on. Once he trusts my intentions, I can move to his hip joints.

If the hip joints are tight and stuck, energy will not flow into the pelvic area (chakra II, sexuality and power) and the back (chakra III, emotions). Energetically, it is vital that the hip joints release their tensions so that this life force can move up the spine, through the torso and neck and out the top of the head. Yet here was Jonathan, his energy trapped in his ankles. Jonathan told me he believed in spiritual practices, but he was cut off at the base of his journey, before it could really begin. His spirituality was an idea in his head, not grounded in life's experience.

I want to discover if he realizes how tight his hip joints are. "Are you aware of your hip joints?" I ask him before I touch them.

"No . . . I kind of know where they are, but I don't feel them."

Most people know the approximate anatomical location of their joints but are not aware if they are relaxed or tight. So I ask him to exaggerate the tightness of his hip joints. He clenches his teeth, holds his breath and pushes his pelvis farther down into the table. Suddenly he relaxes the tensions, breathes out, lifts his head and looks down at his feet. I move his feet again; his energy is still stuck around the ankles.

I slip one hand under his left hip joint and place the other on top of the joint. This kindles a strong sensory awakening: that his pelvis and hip joints are part of his whole self and that he has been denying this most of his life.

"What would you like to say to your pelvis?" I ask.

"You're okay. Just stay the way you are!"

As Jonathan speaks, I feel a dramatic shift in his left hip joint. It softens, relaxes, vibrates. I gather this energetic buildup and with my hands follow it down his leg and out his foot. Eyes moistened, he repeats the sentence, remembering all the times his family did not accept him. Yearning to hear "You're okay the way you are" as a child has fueled his strong need to be accepted as an adult.

My first touch of his hip joint begins a domino effect throughout his entire physical and emotional body. When his body hears the congruency of my words and touch, the texture of his muscles, joints, ligaments and bone shift like harp strings, vibrating together.

"Can you describe how your left leg and right leg feel?"

"The left feels longer, more fluid and energetic. The right one feels stubborn and stuck, like a stick-in-the-mud." I haven't touched his right hip joint yet.

These two legs have become a metaphor for two aspects of Jonathan's personality. One side of him is fluid, wants to dance and move quickly. The other side doesn't want to move at all; it struggles and stays stuck. An inner war is acted out in all areas of his personal and professional life.

I get an image of a man who struggles with contradictory ways of dealing with his parents, friends and colleagues. Each decision keys a crisis: whether to stay put or move quickly. It seems to be time for Jonathan to listen and get to know his two sides. I slide my hands under his left shoulder blade, wait and listen. I feel a tight band around his chest.

"Is there anything you want to say to your stubborn side?"

"Yes. Chill out!"

His chest tightens even more and his breathing becomes shallow.

"Chill out!" he says again, his voice stern. "Like you better do what I tell you or else!"

His lower back begins to arch as he tries to hold back a rush of emotions.

"What does your left leg say?" I ask.

"Everything's going to be all right," he replies in a soft tone. His lower back relaxes its tensions, softens and flattens out on the table, sending energy up to his chest.

"Anything else?"

"Yes . . . I wish my father had said those words to me when I was a kid."

With my hands still under his left shoulder blade, I feel energy moving up his torso, filling his shoulder and spilling down through his arms and out his fingertips. I move my hands along the same pathway, helping him release any residual tensions. His chest opens more fully and he begins to cry silently. I place my hands an inch or so above his stomach (chakra III), moving it lightly upward from his emotional center to his throat, over his face and hair. We are both silent.

I face the crown of his head (chakra VII, unconditional regard and love) and slide my hands underneath it, cradling his head. This is the first time I have the full weight of his head in the palms of my hands.

"Say what you would have liked to hear from your father when you were a little boy."

Jonathan takes a deep breath, clears his throat and says warmly, "I love you, son. I'm glad you're in my life." Again, the theme of acceptance. His previously taut neck muscles are now relaxed and as soft as melting butter.

Touch: The Pathway to the Past

Touch connects us to experiences from the past. First there is a sensory awakening that travels via the nerves to the spine and up the spine to the brain. As the whole body and brain remember early touching, it is the neocortex that identifies the emotions with words.

Emotions and energy are like waves, building momentum, rising to a peak, flowing freely and then ebbing. A gentle, loving touch, accompanied by congruent talk, melts a tight and holding area of the physical, energetic and emotional body.

My first touch helped release Jonathan's emotional and energetic issues into the whole of his body. Subsequent touch and talk allowed his brain—more accurately, his neocortex—to recognize and give voice to his issues. Touch and talk are at the foundation of Rubenfeld Synergy, and the first touch is where it all starts.

EXERCISE: USING THE FIRST TOUCH TO LISTEN (DUET)

When you use the first touch as a means of contacting and listening, you begin the healing process for yourself and others. In the traditional medical model, it is important to diagnose an illness and then cure it. When you touch people from this perspective, your intention is to find out what's wrong, and your touch/contact will transmit this intention. Listening touch, on the other hand, is not nearly this specific. It is trying to discern energy flow and emotion; it is making spiritual and emotional, not medical, contact. No preconceptions exist; no judgments are ever made.

In the following exercise, you will have an opportunity to experience and compare a fixing touch and a listening touch.

Invite a partner who is willing to take enough time to do both parts of the exercise and process what emerges. We will call your partner Toucher/Listener (B) and you will be the Storyteller (A). Find a quiet place where you will not be disturbed.

1. *Storyteller* (A), stand quietly with your knees slightly bent and your eyes closed. Take a few breaths and let your mind go on an inward journey throughout your body, noticing any part of it that is aching for attention.

2. *Toucher/Listener* (B), face A's left shoulder. Make sure you are looking directly into your partner's left ear. Walk toward A until you are close enough to make easy touch contact. Bend your knees slightly and rub your hands together.

3. B, look at A's left shoulder with an attitude of "What's wrong with this shoulder? Is it too high? Too low? Too forward? Too far back?" With an intention of "fixing and correcting," move your partner's shoulder until you feel certain that your remedy has improved the way the shoulder looks.

4. Pause. B, move your hands away from A and shake them out, letting go of the "being fixed" message you have received from your body and brain. A, open your eyes and face your partner. Share anything you wish about this experience.

Questions to Guide You:

For A:

- How did you experience B's touch?
- What message did you receive from B?
- Did you feel "heard"?

For B:

- Did you feel responsible for correcting A's shoulder?
- Did you want to do this "fixing" slowly or quickly?
- How did you feel when you looked at A's shoulder with a critical intention?

5. A, shake out your body, letting go of the "being fixed" message, and return to your quiet stance. Reenter your inner journey.

6. B, face A's left shoulder once again, making sure you are looking into your partner's left ear. Walk toward A until you are close enough to easily make touch contact. Rub your hands together. Imagine "butterfly hands" as you approach A's left shoulder. Move your right hand toward the shoulder blade, passing through the energy field, and gently touch the back of A's shoulder. Let your hand be there in silence—no need to rub, stroke or pat. Move your left hand toward the front of A's left shoulder, passing through the energy field and touching the front of the shoulder. Let your hand be there in silence. Empty your head of any expectations and allow yourself to be fully present in the here and now, listening to your partner. Your first touch makes contact with a whole person, not just a shoulder. In this moment there is no right or wrong. There just *is.*

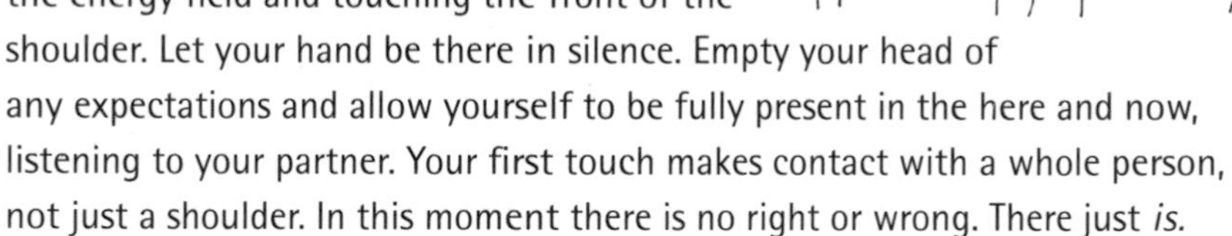

7. Meanwhile, A, continue on your inner journey. Notice where you stop in your body and listen to that part of you. Perhaps there is a situation you are holding; perhaps there's a person with whom you have unfinished business; perhaps a feeling or emotion you've been carrying; perhaps a painful or joyful memory. Who knows? You may just want this awareness, or you may want the situation (or person) to leave your body. You can use your left shoulder as an exit if you wish—or any other place in yourself.

8. Although B does not know what A's story is about, he/she may be sensing a variety of states, stories and sensations in A. B, you are giving a wonderful gift to A at this moment, being totally present and focused on listening to whatever is happening through your first touch. After a while, B, gather whatever energy has been released around the left shoulder. With both hands, move the energy down A's arm, touching the arm, then passing through the elbow and out the fingertips. Shake your hands out, because any energy that flowed through is A's and you do not need to absorb it (an issue of self-care).

9. A, you are still standing quietly with your eyes closed, experiencing any changes in your body/mind.

10. B, walk around and face your partner so that when A's eyes open, you will be there to greet him/her.

11. A, slowly open your eyes and let your gaze take in the person in front of you.

12. B, gaze softly at A and take note of any differences between the two shoulders or anything else you see.

Share anything you want about what happened to you both during this part of the exercise. While it is seemingly simple, the results are often powerful and moving if both partners are focused—so great an effect from so simple an action as a first touch!

Questions to Guide You:

For A:

- Where did you go on your inner journey?
- How did you experience the first touch on your shoulder?
- Did your partner's touch support you on your journey?
- Did you use your left shoulder as an "exit," or another place?
- Did you feel any difference between your right and left sides?

For B:

- Was the image of "butterfly hands" useful?
- What did you experience when you first touched your partner?

- Were you able to sustain your focus on listening to A without wondering about an outcome?
- How did you experience gathering and moving A's energy down his/her arm?
- When you faced your partner, what did you notice about the eyes, shoulders, arms, etc.?

If you wish, you can write about the experience in your journals—prose or poetry. And take some Magic Markers or crayons and draw an image of how you feel. Try drawing your body image before the exercise and then another one afterward. It's fascinating to see what emerges!

The Exercise Continued

By this time, A's right shoulder may be yelling for attention. So B, walk over and face A's right side, making sure you are looking into A's right ear. Repeat Steps 5 through 12, only from the opposite side. And finally, reverse roles. A becomes B and B becomes A, and the entire process is repeated.

Throughout the exercise, notice how much attention the body requires and what it "says" in response to touch. For touching as a therapeutic instrument is not merely used to "feel"; it is used to listen and to hear. What you are experiencing is one of the "Ahas" fundamental to Rubenfeld Synergy.

That the body tells the truth.

6.

The Body Tells the Truth®

Melanie, a petite woman of fifty, comes to my office for a consultation. She complains that she has difficulty breathing, though I do not observe it. I touch her head lightly, go to her feet and then her hips and finally slip both hands under her shoulders. They are soft and malleable. To my surprise, she begins to sob and moan.

"What's happening right now?" I ask.

"My husband's gone. Ran off with his secretary. Left me bereft and miserable. And I thought we were going to spend our golden years together."

"And you loved him?"

"Oh, yes!"

"When did he leave you?"

"A few months ago."

Her shoulders are still relaxed. I check out her legs, stomach and back. Amazing! Her emotional aura does not feel sad or betrayed. In fact, it's pulsating evenly; it does not correlate with her words.

"I'm puzzled. You talk about your husband's leaving you suddenly, yet your body's very relaxed, very vibrant. Are you all alone now?"

There is a long pause. She opens her eyes and grins conspiratorially. "Not precisely alone."

"Then there's someone else?"

She laughs. "Not just some*one.* Two! I have two boyfriends. Both of them give me so much attention, and they're fabulous lovers. I never knew how beautiful lovemaking could be."

I join her laughter. "Then why were you sounding so miserable before?"

Catching her breath, she looks at me, suddenly perplexed. "My husband left me and that's why I came to see you. I should feel miserable and lost. Right?"

Should. How often we think we should feel a certain way—out of obligation—listening to our brain and not our heart, body and soul.

Melanie believed that she should be miserable in her situation, yet she was really enjoying herself. Was she "lying," then, when she told me of her sorrow? Yes and no. Her thoughts were sorrowful but her body and heart were joyful. She wasn't trying to "fool" me—she was really fooling herself. And she would have gotten away with the "miserable" version of her story had she not lain down on my table and let my listening hands hear the truth.

The Mind Remembers; So Does the Body

Our bodies and brains house all of life's experiences. We may not be able to remember them, but they are imprinted in our unconscious. They are there, as much a part of us as our bones, heart and bloodstream. And, whether they're pleasurable or traumatic, dramatic or insignificant, they become part of our character forever. They shape the way we think, feel and react.

We've experienced and seen how the body and mind are interconnected, but they may function differently at various times. The body is a repository of all experiences; it does not censor. The mind is a reflective, thinking tool, capable of changing and molding experience (depending on the situation) to its wishes. You can change your mind and talk yourself out of an idea and feeling. You can think about an event one way on Tuesday and view it entirely differently on Wednesday. The mind can dramatize, romanticize, make the unhappy bearable, the painful justifiable and the sorrowful joyous.

Claire's shoulders and chest began to collapse when she was a little girl. Throughout her childhood, her mother was cold, distant, uncaring. "It's because she had to work so hard, keeping me clothed and well-fed, and did not have time to show me much affection," Claire rationalized, explaining the emptiness in her stomach just after her mother died. "I know she loved me all the same. I know what happened and I understand. I forgive her." This is what Claire thinks and *wants* to believe, yet she continues to walk with rounded shoulders and finds it hard to look people in the eye. When Claire sees a mother holding and kissing her child, she feels a shot of pain in her gut, a tinge of discomfort around her heart. Her body remembers the lonely, cold hours without her mother and *it* hasn't forgiven her mother yet.

You can stretch, exercise and even surgically alter your body, but it will continue to tell the truth, more and more forcefully, until you listen. The body has memories stored in every cell, bone, nerve and muscle. My analyst friends were surprised when I first used the expression "cellular memory." But now they acknowledge that talk alone may not deal with those memories that are lodged deeply in every molecule. One of the most powerful tools to contact cellular memories and engage the limbic brain is touch.

The skin remembers touch, or the lack of it; the nose remembers smell, the ears remember sound and music, the mouth remembers taste. And so the taste and texture of Proust's famous madeleine can evoke memories not only of place but of feelings—not through the mind but through the senses. Although the mind can shape memories with words—in Proust's case, he shaped them into art—they were originally a deeper truth. They were body-memories first, uncensored and unformed.

Interconnections

To try to separate the mind and body is an impossible task, though over the centuries both philosophers and scientists have tried. Because they are so interrelated, all sorts of complexities immediately arise if we try to treat their functions separately. You may have heard of the "phantom limb" phenomenon—when people lose part of their body, but their brain retains the template of that part. I remember a man who still experienced itching in his right foot although it no longer physically existed.

Another fascinating phenomenon is "phantom pain." A psychiatrist told me about a female patient who suffered with crippling arthritis in her knees. For years she denied the pain and refused to do anything about it except take some painkillers. Eventually she followed her doctor's advice to have knee replacement surgery. Recovery took some time, which required her to exercise, walk and use her new artificial joints. Anatomically, everything was in place and functioning. The pain of the operation receded. However, the pain in her knees remained. The psychiatrist explained that if you touched her knee lightly she would wince in pain, even though the neural pathway from knee to brain no longer existed. Her *brain* retained the former pain. So just as the body remembers events the brain has forgotten (or altered or repressed), the brain remembers the body's parts and the sensations they felt. The body and brain, the present and the past, interact in fascinating and therapeutic ways.

Ruth, a young forty, has lost her right breast to cancer. She comes to my workshop, telling us she's worried about a career change. When she lies on her back, I can see she has not used any prosthetic device to match her other breast. Saying nothing, I gently touch her head. A wave of intense energy streams from her second chakra and pours into the vessel of the third, the emotional center. Almost immediately she begins to cry.

"Is this sadness about a job choice?" I ask.

"No."

"Then what?"

"I've been so sick. I lost my breast and I'm grieving for it. It was part of me for thirty years, and now I'm unbalanced and no longer whole."

I slide one hand under her back and leave the other on her left shoulder. I see her breath rapidly cresting.

"It's gone forever," she continues. "It's betrayed me."

Her tears increase, and I can feel her tightness as she tries to stem them. "Let the tears go," I encourage her. "It's all right to cry."

Then I turn to a technique I've used successfully in thousands of cases of loss: a dialogue with the missing part. "Talk to your breast. What do you want to say?"

Her eyes flash. "That I'm angry."

"Imagine your breast answering. What would it say to you?"

" 'I'm sorry.' "

Her tears become a torrent, intensifying her emotional aura. I move her energy and assist it up through her chest. "Let your cries flow naturally, like the ocean's waves," I tell her. "Nothing can stop waves. They splash into the sand and then move out again." She takes a deep breath, and the release in her body is palpable.

"Now let your breast talk to you again. Before, it said it was sorry. Is that all?"

"No." Her head shakes from side to side. She touches the place where her breast had been. "It says, 'I didn't want you to be hurt. I loved you very much.' " Ruth needs no further prompting from me. "And I say to it, 'I miss you. At first I was angry at you for making me so sick, but I love you—and oh, God, I'm lost without you!' "

"Isn't it wonderful that your breast sounded the alarm and sacrificed itself?" I ask her, my tone quiet but penetrating.

She peers at me in wonder, eyes widening. "Yes! My breast is saying, 'I'm glad you're alive.' *That's* what it's telling me. It's happy for me. Happy for the gift of life it gave me."

"You might want to thank it," I say softly, smiling.

There is a long pause while she integrates this new way of thinking and feeling. Then her tears turn into laughter. "Thank you!" she exclaims. "Oh, thank you!"

Ruth is so engrossed in this process that she has forgotten there are other people in the workshop. Slowly, as she sits up, her spine lengthens and she peers around the room. Now I also focus on the group. Many in the room are crying, surely coming in touch with their own losses. The universal themes of loss and grief touch everyone. It need not be the loss of a body part; it can be the loss of friends or family, or the loss of strength, appearance or ease of functioning.

EXERCISE: A VISUAL JOURNEY (SOLO)

Having a dialogue is not the only way to process loss. Another way is through drawing. Take a few minutes to try this exercise:

1. Find a time and place where you will not be disturbed, taking with you several sheets of blank paper with crayons and/or Magic Markers. Sit quietly and take a few deep breaths. Let your mind and body travel back to a time before your loss.

2. Open your eyes and without thinking too much draw your body before the loss, using any colors you wish; the picture does not have to be "artistic" or realistic. Include some colors that represent your feelings and energy.

3. Now simply look at your drawing and allow any emotions to bubble up and move through your body. Then take a few more breaths and close your eyes.

4. Open your eyes again and take another piece of paper. Draw the part of you that's missing or isn't functioning. Again, use all the colors you want and express your feelings visually. Take your time. You may want to express your emotions or just share your thoughts through the drawing. No one is around to criticize or judge you.

5. Place both drawings in front of you and look at them. Let your eyes move from one to the other. Notice your breath.

6. Using another blank sheet, draw yourself as you experience your body now. Without planning, use the colors that "feel right." Remember: this is not going to be judged; it is for you and you alone. When you finish this third page, rest a few moments.

7. Take all three sheets and place them next to each other. Before you is a visual journey through your physical and emotional state before and after the changes. You can see how you viewed yourself at various times in this experience.

8. Finish this exercise by saying anything you want to the three drawings. Then imagine golden healing light around your third drawing. This is how you are now. If there is any part of you at this time that still needs healing, it will call on you, if you take the time to listen.

EXERCISE: JOURNAL OF LOSS (SOLO)

Yet another way to deal with loss is by keeping a journal, as in this exercise:

1. Once again find a quiet space so no one will disturb you. Have a notebook handy. Have a pen, pencil and coloring pens at your side.

2. As you take a few deep breaths, let your mind and body travel back to the time just before your loss. Allow your thoughts to flow in and out of your mind without judging them—as if you are watching pictures pass by.

3. Open your eyes and in your notebook begin to write anything you wish about how you felt during the experience.

Questions to Guide You:

- How did you discover there was a problem?
- Did anyone tell you or did you sense that something was not right?
- If a medical practitioner told you, how did he/she share the information?
- Did you have a support system in place to help you?
- Did friends or relatives help you through the shock and disbelief that occur in the first stages?
- Who was there to share your fears and uncertainties?
- How did you respond to suggestions from others?
- What were your feelings when others shared their emotions?
- How did you respond when someone said, "Oh . . . you'll be all right," when you were in the middle of the realization that you were *not* all right?

These questions are meant only as guidelines to assist you in writing about the incident. Your own questions can and should be more specifically related to your experience.

4. Once you've completed writing about this first phase, stop and rest. Breathe deeply and close your eyes.

5. Now open your eyes and turn to a blank sheet. Write as though you yourself were the "loss," letting it speak to you about its experience. Allow the "loss" to say anything it wishes, without censoring any thoughts, feelings or associations. Complete this part. Close your eyes and rest.

6. Open your eyes and turn to another blank sheet. Divide the page in half. On the left side will be the voice of your body before the event, on the right the voice of the part (or person) that was lost. (For Ruth, the left side would be her whole body, the right column her breast.)

7. Now, write a dialogue and let these two columns speak to each other. Once you have written everything you feel, take a moment to rest and reflect.

8. Review what you have written and notice how and where your body reacts. Give this time so that you can integrate all this material physically, emotionally and energetically.

9. In this last part, turn to a blank sheet and write down how you are feeling right now.

Questions to Guide You:

- What is your body image at this time?
- What lessons did you learn from the loss?
- How has the loss changed your life view?
- What helped in bringing your life and body to a whole place?

The exercises in this chapter have been designed to help you confront, express and listen to such emotions as grief and loss. In Ruth's case, the verbal dialogue with her lost breast (while she touched her chest) enabled her to express her grief, release the emotional pain housed in her body and, in effect, come to terms with this chapter in her life. Certainly, working with an experienced Synergist assisted her in the process. However, the exercises demonstrate that there is much you can do to deal with these themes on your own and/or with a partner.

Listening to your body's story demands a commitment of time and a willingness to take a journey inward. Once you have learned how to travel inside and hear messages from parts of your body, you can add touch to heighten the body's truth and wisdom. In every story, the body part usually becomes a metaphor for other life issues. As Ruth confirmed later during ongoing sessions, she discovered that the theme of loss was strongly rooted in her family history and her childhood.

Double Messages: Incongruencies in Everyday Life

"You've been with us for twenty years and your work has been very good," a boss tells a manager with a smile. "However, we've been bought by another company and we're downsizing. Sorry about this, but I'm going to have to let you go."

Bad news ("I'm firing you") has been delivered with a smile. These two messages are not congruent and the emotional impact is confusing. The manager receiving the "double message" has to deal intellectually with the content and consequence of what he is being told and at the same time experience the import energetically and nonverbally—in his body.

And the *smile!* What's going on?

- The boss may be trying to get the employee to mirror his smile, thinking that the employee then won't feel so bad about being fired.
- The boss may be trying to avoid his own pain, hoping that his lighthearted expression will bring him inner well-being.
- The boss may be *unaware* of smiling while delivering his devastating pronouncement—and so out of touch with his feelings that his facial expression has no relation to the information he's imparting.

In each case, the boss is expressing his covert, unspoken feelings, which transform themselves into an undercurrent of energy that is internalized by the receiver's body. Many clients have reported that there are times when a particular message should bring them joy and happiness, but they instead experience pain and discomfort in their stomachs. The messages from their bodies are not congruent with what is being said. They are experiencing a "double message" in a clear physical way.

Time after time we adopt a demeanor that does not reflect our inner feelings. Indeed, when we continually *direct* our body to respond in opposition to our true feelings, these feelings may change, sometimes to our benefit. Adopt an assertive stance, for example, and fear may change to confidence.

A friend of mine wrote advertising copy for an unpleasant boss. Although he was well-paid, he did not appreciate the tensions and stress at his job. When he found out that a particular assignment had been awarded to another writer, he expressed his anger and disappointment to me.

"But you look relaxed," I said, puzzled. "How does your body feel?"

"Terrific!" he said, astonished.

He was relieved, not disappointed, that the job had been given to someone else. Like Melanie, he was *supposed* to feel angry and disappointed; his body told the truth.

The Truthful Body and the Complex Brain

Our brain, according to Nobel laureate Paul MacLean, is made up of three main sections (called the *triune brain*). Its top part was the last formed—over three million years ago. Called the neocortex, it is the seat of rationality. It organizes and processes information in a practical and theoretical context; it wants to know *why* and *what.* What is history? Why does two and two make four? What does Mary mean when she says, "I love him"? Why does she love him in the first place? It is able to communicate in words and symbols—reading, writing and 'rithmetic are its tools. Without my neocortex, I couldn't write this book. Without yours, you couldn't understand it.

The limbic part of the brain is much older, probably going back a billion years. It houses all nonverbal experiences and events. It is our emotional center, reached through the five senses. It responds to music, smell, sight, taste, movement, art and touch. It is nonrational, yet it is capable of remembering. We are affected by it viscerally. It's why we grow teary-eyed at rehearing a piece of music experienced long ago with a lover, why we salivate at the smell of a well-cooked dish, why we *care.*

Finally, there's the reptilian brain—the oldest of the three parts, its origins maybe two billion years in the past—located just below the limbic center at the base of the skull. It is our flight-or-fight mechanism; it enables us to survive. The reptilian brain responds to threatening situations quickly, nonrationally, and it is so tied in to our limbic brain that the two sections are sometimes referred to as the limbic system. (I once called my workshop "Out on the Limbic," but too few "got it" in their neocortexes, so I stopped.)

The body is most closely connected to the limbic system because it, too, is nonverbal and nonrational. It remembers childhood family scenes, for example, and translates them into bodily sensations and postures which become life metaphors. Thus, when you walk with stooped shoulders or on tiptoes, your body is telling the truth about past experience which your neocortex could not put into words, and indeed may have forgotten.

Jerry: Imprints of the Past

When I was growing up, I spent some time at my neighbors' apartment playing with their son Jerry. He loved to sit on his father's knee and be kissed and cuddled. On some days, his father did just that. Unfortunately, however, his father was an alcoholic, and sometimes he came home in a fury. Jerry, expecting loving touch, got shouts and even fists. I saw little Jerry begin to shrink away with fear when his father arrived, not trusting him anymore. After all, disaster could and did strike without warning; nothing in his world was predictable.

We both grew up. Jerry's father died. Now when we met, I was shocked to see this good man standing and walking with shoulders hunched and head down, cowering. Once in a while, with difficulty, he would look me in the eye, but mostly as an adult he was still a little boy, avoiding human contact in order to survive. He had developed a self-defeating preparation against imaginary attacks and so shied away from all people, severely crippling his life. Rationally—in his neocortex—he knows that not everyone is his father or an alcoholic, but to his body, the traumatic truth of the past is more real than the rational truth of the present.

Jerry's physical patterns hold the emotional history of his early life. He needs to be reached through touch and movement (his limbic system), not talk alone (his neocortex). The imprints of past events are locked in his body, and it is there he must change to overcome his fears. Psychotherapists and researchers such as Dr. Bessel van der Kolk and Dr. Francine Shapiro have discovered that severe trauma cannot be treated or changed through verbal intervention alone.

Elizabeth: From Trauma to Recovery

Once in a lifetime, if you're lucky, you get the opportunity to meet a person who has survived experiences almost beyond human imagination, and to hear her story through listening hands.

At a workshop some years ago, a woman named Elizabeth immediately caught my attention. She wore a black patch over one eye, and her hair was thinning, reminding me of people undergoing chemotherapy. She smacked her lips as she watched me demonstrate Rubenfeld Synergy, and with a special urgency she volunteered to be the next client.

I asked her to tell me a little bit about what was going on with her life. She told me that her great passion in life had been to become a reconstructive surgeon, because of a pivotal experience at age three. Her horse, Jasmine, was her "special friend," and she would pet and stroke it every day. But the horse broke its leg, and Elizabeth's father took her to the pasture with him and shot Jasmine, ending the horse's agony. It was such a traumatic moment for Elizabeth that the little girl made up her mind, heart and soul to learn how to repair broken limbs.

Elizabeth graduated from medical school as a reconstructive surgeon and practiced for many years. One day, however, she woke up with double vision and gradually had to face the reality that she could no longer continue her profession. Over the years she had traveled to Africa and to South and Central America, both as a healer of limbs and a teacher of native doctors, so that they could perform some of the "miracles" they saw her do.

At this point, I asked her to lie down, and explained that she could stop the session for any reason whatsoever. When I passed my hands over her body, I made contact with an exceptionally dense energy field. I gently touched the sides of her head and began to sense her fear. Her skull and neck were tense and she did not want to move. I went to her feet. They were large and solid, creating a strong foundation for her body. When I touched her shoulder, her chest quivered, as if to say, "I don't want to be relaxed! I need to stay on the ground and not fall apart." The texture of her muscles was more than tight; the muscles were hard, forming a defense against anybody or anything that might intrude. My hands tingled for a long time, warning me to be extremely careful.

After a long, quiet pause, I asked her what she was experiencing.

"Your touch is loving and gentle," she said. "Yet I get so frightened." Her eyes fluttered. "I went back to a small village in Peru where I was teaching the medical practitioners reconstructive surgery and spending hours helping them repair the faces and limbs of little children. Since I was the only American woman, it took them a long time to accept me, yet eventually a deep bond developed between the villagers and me.

"One day, a band of terrorists attacked and threatened to kill the hospital staff. I begged them to spare my students and to take me instead, and to my misfortune they complied—I was the enemy, for sharing 'Western secrets.' Isolating me, they tortured me mercilessly, threatening to kill me. 'Today is your last day,' they screamed.

"Then—a miracle! Soldiers appeared one day and one of them said, 'Mamacita, you're free to go.' I was wounded, weakened and starved. They picked up my aching body and dumped me in the village square, then disappeared into the jungle. The villagers, many of whom I'd helped, carried me to one of their homes, where they nursed me through a difficult time. As soon as I was well enough to travel, the American Consulate met me and arranged a flight back to the States."

Elizabeth told this story haltingly. She had somewhat recovered physically, but the emotional trauma to her body and brain was indelibly printed in every cell.

Now I understood the terror in her body. My fingers touched her lightly, reminding her that I would not hurt her and was still present. "What kept you going?" I asked.

"I prayed to God and wrote poetry."

"Would you share one of your poems with us?"

She took a deep breath.

"I see the pain take wing and fly
And see the clouds which dribble the rain in healing embrace.
Where sutures once were,
There is tenderness and grace."

When Elizabeth was finished, she opened her eyes, stared into mine and smiled.

"What was the funniest and most outrageous thing that happened to you in that village?" I asked, knowing intuitively that even with all the pain there must have been humorous moments.

"I haven't thought of this for so long," she answered gratefully. "I have big feet, especially compared with the villagers and their children. When I was recovering, each morning the little ones would hide my shoes. It became a ritual between us. When I found them, they'd howl with laughter. Those children were my special loves. Now I visit hospitals dressed as a clown to entertain the sick children—it's the least and the most I can do."

Laughing, she sat up and looked around at the group. Slowly, I guided her off the table, and with her feet firmly planted on the floor she stood still for a few moments, then made her way to a chair and sat down.

A heated discussion followed about justice, cruelty and betrayal. But this came from our rational brains. What was evident in everyone's posture was love and compassion, which they silently communicated to Elizabeth.

My own hands still remember her body's truth.

TALK AND TOUCH

After listening to many people's body stories, I am convinced that it is the *simultaneity* of touch and talk—therapy for the limbic system and the neocortex—that leads to lasting change. Dr. Candace Pert's research has shown that receptors for the molecules of emotion are housed in our blood, in our bones, in our muscles, even in cells of our digestive tract and immune system, and that touch or massage alone may elicit an emotional outpouring. Talk alone may help clients remember and understand the reasons for their dysfunctional behavior. But Rubenfeld Synergy engages both the body and the brain. If they are congruent, emotional truth will emerge.

Mental stress affects the body; bodily pain affects the brain. Ignoring signals of pain and stress creates denial—Maybe if I ignore the signals, they'll disappear, we think. But life teaches us the opposite. Stress and pain don't disappear; indeed, they only intensify their demand for attention, shouting at us until we notice.

I am so used to handling many demands and challenging situations that sometimes my will to forge ahead drowns out my pain's insistence on being heard. The results may be devastating and eventually manifest themselves in all sorts of illnesses.

In my late twenties, I was frantically preparing and organizing a program of choral and orchestral compositions for my debut at Carnegie Hall as a conductor. I couldn't afford a manager or a production crew, so I oversaw every detail myself, from designing the program, printing and selling tickets, marketing and hiring the musicians to rehearsing and performing the concert. This was an extremely stressful time. I slept little, practiced for hours every day, and worried. A few weeks before the concert, my father died suddenly. My shock and grief exacerbated this already difficult time.

Superwoman—yours truly—was, I thought, up to the task. I handled the funeral arrangements and took care of my mother, all the time continuing rehearsals. I rededicated the concert to my father.

My right arm began to hurt.

I paid no attention to the first twinges and completed the concert. Soon after I developed a full-blown tendinitis in my arm, making it impossible for me to conduct, write or work with my students.

I can recall a far more dramatic case of brain influencing body.

Judy, a young-looking forty, arrives, hunched over as though she has osteoporosis. Her hair is matted and unkempt. In fact, Judy has bone cancer. She has already undergone several sessions of chemotherapy and has received a bone marrow transplant. Now she is about to undergo yet another round of chemotherapy. Her physical condition puzzles and distresses her. She shakes her head from side to side, muttering, "Why?" Yet this is only part of what's troubling her.

"Nobody loves me," she says.

"Really? What about Alice, your friend who recommended me? She told me that you're wonderful. Seems to me she wouldn't be so involved or concerned if she didn't love you."

Her frown momentarily disappears. "Alice adores me. She's an exception."

"Oh. Are there others who love you?"

"There's Joseph," she admits. "Joseph and Sarah."

"Family members?"

"No. Friends." She begins to cry.

"Can you think of anyone else who cares about you?"

She mentions three more. The pattern is becoming clear. It is not that Judy is unloved, but that she doesn't love herself. In the session she had discovered several facets of herself: the unloved Judy, the worrying Judy, the unworthy Judy. How could any of these Judys possibly attract a partner? she'd asked herself. She'd remembered deciding at age twenty that if she wasn't married by the time she was forty, she'd kill herself.

And now, at forty, she has cancer. I slip my hands under her back and cradle her shoulders. I point out the correlation between her unmarried state, her age and the decision she made. Judy "gets it." She understands that her body is sending her a vital message—to love herself. It is demanding attention in the most metaphorically dramatic way possible.

She says, "You know, marrow reproduces life. I've always felt unloved, and that's why my marrow is diseased. It's death." Literally, her anxiety, her feeling of aloneness, her morbidity was in her bones.

"Six people care about you," I tell her. "Alice adores you." Speaking softly, I repeat these assurances again and again.

Her body softens; I can feel her pleasure at my words. It is too soon yet to tell if she will come through her ordeals. But at least for the moment, I am able to support her and help her to relax. If she is able to be more open to affection and love, her body may open to healing in turn.

Posture Affects Emotion, Emotion Affects Posture

Nowhere is the correlation between body, mind and emotion more apparent than in body posture. Some of the way you walk and stand is dictated by genetics, but it is also greatly affected by how you feel—by emotions of the present and deep-seated emotions of the past. As you become more aware of your posture, and more willing to explore your embodied feelings, you will discover an important key to aligning yourself from the inside out. Just as the mind needs to be aligned with the body, so the body needs to be aligned with itself, and posture is the means to change.

At the conclusion of her Rubenfeld Synergy training, one of my students, Margaret Tucker, presented her certification project on how children develop their posture. Through a series of photographs of infants at various stages of development—crawling, sitting, standing and walking—she demonstrated how genes, body structure and emotions all influence posture, even at such an early age.

Small children inherit their parents' genetic posture and imitate their parents' emotional one. Even the emotional messages unborn children receive in utero may have a powerful impact on their posture and health.

During my mother's pregnancy, she was extremely anxious about my father's typhoid fever. This life/death situation clouded all the joy she initially experienced in becoming pregnant. Her nonverbal message of pain and anxiety was not about me, but about him. Yet as a child I experienced free-floating anxiety attacks in my body. My rounded back and tight chest were the consequences of these episodes. It wasn't until I studied the Alexander Technique with Judy Leibowitz, and explored the meaning of these episodes, that I changed my habitual way of standing and moving.

Often, my students and clients seem to change posturally only to return the following weeks with their familiar physical problems. Since old emotions are so deeply embedded in our bodies, the process of postural change is never easy. Yet it is at the heart of the Rubenfeld Synergy Method, for there is no question that postural change leads to emotional healing—and change is assuredly possible.

Your nervous system is the messenger of your mind. It prepares and organizes you to move from the moment you decide on or think of a motion—reaching for a fork or stepping over a puddle. Similarly, the change of your body's position affects your mind. And your emotions are the bridge that connects your mind and body. For long-term change, a heightened awareness of your emotional posture is essential.

So let's begin here with two simple exercises to allow you to focus on your posture. There will be more later in the book.

EXERCISE: FACE TO FACE (DUET)

We talk to other people many times a day, yet we are not usually aware of the habitual use of our body in the process, until it cries out for attention.

1. Find a partner willing to do the exercise with you.

2. Imagine the two of you have just met. Stand in a comfortable position and face each other.

3. Slowly stick your neck out and jut your chin forward. Stand in that position for a few moments and then talk to your partner from that posture. After some time, pause and let your neck move backward and let your chin relax.

Questions to Guide You:

- What did you say when you stuck your neck out?
- What did you experience?
- How did your throat feel?
- What happened to your breathing? To your voice?
- Did any emotions bubble up?
- What happened to your eyes and sight? What about your peripheral vision?
- Could you move your head from side to side?

Variation One

1. Shake everything out and wait a few moments. Once again, imagine the two of you have just met and face your partner.

2. Allow your chin to move down toward your chest—your head will follow. Make eye contact with your partner even though your face is down. Talk to him/her for a few minutes in this stance. Now pause, move your chin up from your chest and bring your face back to center. Take a deep breath and process the experience.

Questions to Guide You:

- What did you experience in your eyes? Did you notice any emotions?
- Where did you feel stress and strain in your body?
- What happened to your breathing? How did your back and spine feel?
- Did you notice any difference in your voice and its quality?
- Were you able to move your head from side to side?

Variation Two

1. Begin again by facing your partner. This time move your chin toward the ceiling while maintaining eye contact with him/her. Wait a few moments and begin talking from this posture.

2. After a few moments, pause, shake everything out and bring your chin back to center. Share everything you wish about your experience with your partner.

Questions to Guide You:

- Did your partner look different?
- What did you notice about the back of your neck and spine?
- Did your neck vertebrae and spine move up or down?
- Did your voice change its quality?
- Where did you feel any stress or strain?
- What emotions emerged?
- Could you move your head from side to side?

You can do these exercises by yourself, without a partner. Use a full-length mirror and talk to yourself, asking whether you recognize the posture. Is it one you know?

EXERCISE: UNIVERSAL POSTURES (SOLO)

Remaining in a certain posture over a length of time may produce distinct emotions, such as anxiety, depression, sadness or joy. Anthropologists have studied the physical manifestations of human emotions in great depth (Darwin devoted much time to the emotional expressions of animals and human beings) and found many to exist in all cultures.

One such universal movement is flinging the arms upward—a stance of joy. Indeed, stand with your feet apart, fling your arms up and say, "I'm depressed!" You'll laugh, I promise you. Now try casting down your head and eyes and saying, "I feel great!" This won't work, either.

While posture can affect emotions, the reverse is also true. Try this exercise to learn how your body reacts to emotions:

1. Stand in a "neutral" (centered) position, your arms hanging at your side and your head comfortably balanced.

2. Close your eyes and imagine a time when you were full of *fear,* or think of a scene that evokes fear. Allow your body to move into a posture of reaction to fear. Hold it for a few moments, then exaggerate it and make any sounds you wish.

What are you aware of? Your breathing? Your facial expression? Your posture? Your eyes, mouth or nose? Tensions and muscle holdings?

3. Now shake the "fear" out, return to your neutral stance, take a deep breath and rest. Write anything you wish about fear.

The same exercise can be used for *anger,* for *sadness,* for *joy* or *happiness,* indeed for any emotion you feel strongly at any given moment. In your journal, note your reactions to each emotion after the exercise is completed.

Awareness of your body is one of the key elements of change—psychological or behavioral change is far more difficult without it. As we'll see, other elements are involved, particularly the therapeutic trance. Yet there are a number of people who don't think fundamental change is possible, who believe that we are locked into our characters from the moment we are born or by the time we reach adolescence. My practice—my entire life experience—convinces me they are wrong.

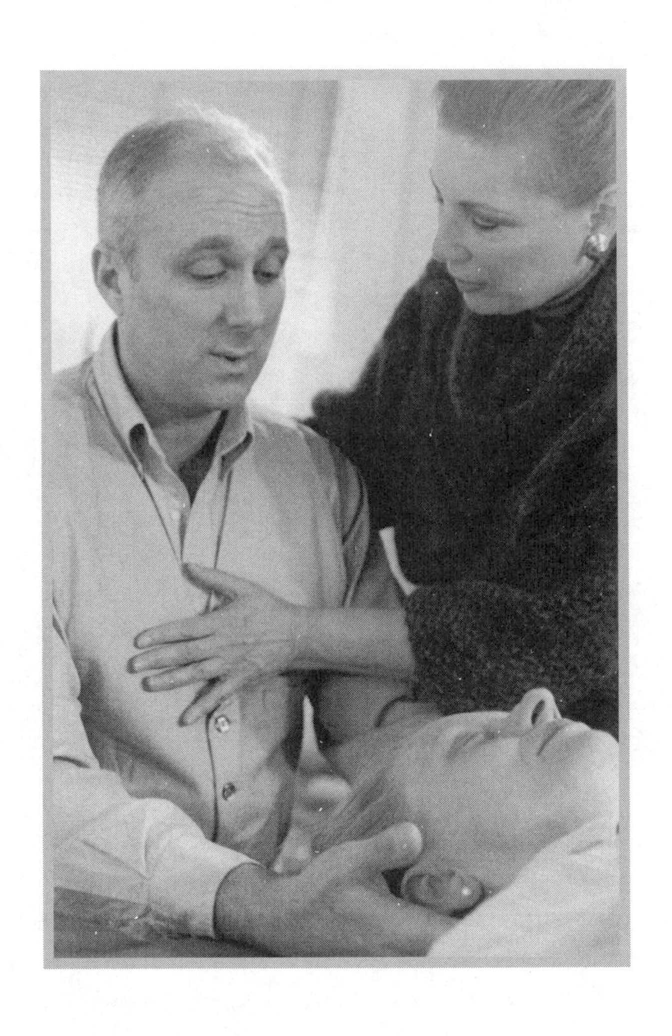

7.

The Capacity to Change

A Buddhist monk is hungry. He goes to a hot dog stand and says, "Make me one with everything." The vendor hands him a frankfurter with all the trimmings. The monk gives him a $20 bill, which the vendor pockets. The monk asks, "Where's my change?" "Ah," replies the vendor. "There is no change from me. Change is within you."

Indeed, the power to change *is* within each of us. Inner self-awareness is the first step in the journey of change.

From Inner Awareness to Changed Behavior

Rubenfeld Synergy is designed to support and listen as you learn to unlock emotions and feelings trapped in your body. Only by first becoming aware of these emotions, then reorganizing, expressing and feeling them fully can you understand your behavior and how it may be limiting your life. It is a process that takes place over time, and it begins with awareness.

I remember my first encounter with Dr. Fritz Perls, psychiatrist and co-founder of Gestalt therapy, at the Esalen Institute in the 1960s. Fritz was seventy years old when I met him, but his reputation for no-nonsense confrontation combined with a puckish sense of humor and Don Juan encounters preceded him. People either hated him or loved him; there was no middle ground. I grew to love him, despite his proclivity for unwanted sexual advances. In our initial training session—we were to be with him for a month—a group of twenty of us, all strangers, sat in a circle around Fritz, expecting him to ask our names, where we were from and what we expected from the training. Instead, he asked each one to complete this sentence:

"Right now, I am aware of——— "

No one had ever asked me so directly what I was aware of in the present moment. Having been taught to be a "good girl," I sat uncomfortably, trying to form the "right" answer. My turn. A very long pause. I could say nothing.

"Exaggerate what you're doing," Fritz instructed.

"Exaggerate what?" I asked, not knowing what he meant.

"What you're doing!"

But what *was* I doing? I first had to become aware of it. I realized that I was sitting rigidly, arms folded across my chest, legs intertwined, staring down, literally holding myself together. I tightened my arms and my legs.

"More!" he declared.

What did he mean?

"Do what you're doing. Tighten yourself more."

The group became blurred. All I heard was my heart racing. Sweat formed on my forehead, and my stomach knotted with anxiety. I pushed harder into the chair, became tighter, could hardly breathe. There was nowhere to go or hide. I reached a place of no return and I felt like a tight ball, full of compressed energy. Suddenly I flung my arms apart, uncrossed my legs, took a deep breath and yelled, "I hate being here!" I turned to Fritz. "And I don't like you. I don't know how I'll survive here for a whole month!"

There was silence. The group, suddenly very clear to me again, was staring at me. I looked at Fritz apprehensively.

He smiled. "You see," he said to the group. "Ilona [he always mispronounced my name] was able to go inside herself and become aware. By exaggerating how she was holding herself, she brought her imploded posture to conscious awareness, and expressed her emotions authentically."

Indeed, Fritz had created an awareness within me that I had never anticipated or imagined. My body was expressing my real feelings ("Help—get me out of here!") while my brain was trying to come up with the "right" answer and make everything okay. In my habitual pattern of withdrawing and keeping silent about my feelings, I suddenly realized, I was repeating what my mother had always done when I was a child.

The group members understood what was happening. They were smiling at me; *they* did not withdraw. On the contrary, they were making contact with me, and it did not take me long to reciprocate.

The Desire to Change

My experience over the past forty years has taught me that people genuinely want to change. They're often unclear whether they want to change their behavior ("How can I be less angry with my children?") or their character ("How can I become a less angry person?"), but they're troubled by some aspect of their life and want to ease their discomfort.

Some people are forced into change, often against their will. Severe trauma, such as the death of a parent or child or a sudden debilitating injury, forces self-examination and with it a reordering of behavior. I suspect that all of us faced with a sudden tragedy vow to "change our priorities," but unless we have outside help or devote the time and effort needed for true change, we probably return to old life patterns.

Many people come to workshops or Rubenfeld Synergy sessions because of terrible emotional and/or physical pain and suffering. "I can't live like this anymore" is a sad and, alas, familiar refrain which fills me with compassion and concern. Then there are those who are not facing life crises but want better ways to deal with their everyday challenges.

Marianne: "Power Surge"

Marianne, a large and spirited woman, eagerly volunteers to be a client in a workshop session. Her energy field is dull. She is in distress because menopause has produced severe discomfort in her body. "Why is this time so difficult, unpredictable and painful?" she asks. When I touch her she tightens, as though flinching from physical pain. Her shoulders and back are stiff.

"Touch a part of your body with your hand," I instruct her.

She immediately places her palm on her forehead.

"What does your forehead say?"

"It hurts. My forehead hurts."

"Can you say 'I hurt' and notice how you feel?"

"I hurt. I hurt." She begins to cry. "I'm too young to be having menopause. My mother was fifty and I'm only forty-four."

"Maybe it's time to ask your mother what menopause was like for her."

"I did. She said it was a piece of cake, and that's not much help."

Her body begins to tremble and her energy field brightens as the pulsations become stronger. "My body's sudden shifts make me cranky. There's no warning. I'm afraid."

I ask her to start a sentence with "I'm afraid because ——— "

"I'm afraid because my situation will affect my ten-year-old daughter," she responds quickly. "I'm afraid because I can't tell my friends and they don't know what's happening to me. I'm afraid because I can't let my discomfort show to my patients. I'm an allergy doctor and these hot flashes happen during consultations."

When she says "my patients," her body tenses markedly and her energy field darkens again. It is telling me the truth. Of all the situations, her fear of exposing her discomfort to her patients concerns her most.

"Are you willing to try an experiment?" I ask intuitively, using a cheerful voice to counterbalance her darkening mood.

"Yes."

I want to go straight to her emotional/sensory processing. One way is to access her visual imagination. "What's your favorite color?"

The answer comes quickly. "Yellow. Like bright daisies."

"Close your eyes. Imagine you are in your office. A hot flash starts and you go over to your desk, pull out a yellow handkerchief embroidered with daisies and a yellow fan and begin to fan your face and wipe your forehead."

She goes through the motions, stops suddenly and declares, "I'm going through a power surge!" She laughs and the group joins her. "But what do I do when I'm seeing a patient?"

I ask for a volunteer who will role-play a patient. She and Marianne sit opposite each other. Marianne explains that the patient is allergic to many substances and that she's sneezing and coughing. The volunteer sneezes and coughs. Marianne says, "Excuse me," walks over to an imaginary desk, whips out an imaginary yellow cloth and announces, "I'm having a power surge."

She waves the cloth around, fanning herself. The volunteer joins her laughter and thanks Marianne for sharing what's happening to her in such a lighthearted way.

I touch the sides of her ribs and back. They are very relaxed. "How do you feel now?"

"Much better." She pauses; I can virtually *see* her coming to self-knowledge. "This is so different from my usual behavior. I worry about what people will think of a doctor who admits to uncertainty." Marianne looks at the warm, receiving faces of the men and women surrounding her. A few comment on how changed she appears.

For she *has* changed. Here is an example of a session that supported immediate change of behavior by transforming a "heavy" problem to a light attitude. Exaggerating where people are often highlights and deepens how they feel. Marianne was already aware of her distress, and therefore lightness was teaching her that a different behavior is possible. She liked what she found. Although Marianne was in distress, she kept her condition a secret. Asking her to reveal her secret, even flaunt it with a big dollop of humor, changed her experience from one of oversensitivity and fearfulness to an honest sharing of those fears and acceptance of herself.

The change process is neither bad nor good. The *I Ching* says: "The only *constant* is change." Change is like breathing, essential for life. Every cell grows, builds up and breaks down. We could not function without this miracle of life.

John: Awareness 101

Even though we express a desire for change, we are not usually aware of our inner physical and emotional battles and are blind to the habitual behavioral patterns ingrained in us from childhood. That's why *awareness is the first step* in any process of change in the body, mind and emotions. Recognizing our *inner resources* is also essential to the change process.

John's shoulders were rounded, his chest sunken, his eyes staring at the floor. "I'm so glad to be here and make contact with you all," he said in a flat voice. His rapid breathing increased and he looked frightened. He was totally unaware of the incongruence between his body language and what he was saying to the group. Indeed, he believed that his words showed he was an open, receptive person. The work he needed was Awareness 101.

I sat facing his right shoulder and put my left palm on his upper back. His spine was rigid in its rounded form.

John complained about not having lasting relationships and wondered why people pulled away from him, despite his efforts to become close.

"What if you made eye contact with someone in this room?" I asked him. He became pale and frightened. Without knowing the details of his life story, I asked him if he would be willing to try an experiment. He agreed when I assured him he could stop at any time.

"Look around the room and ask for a volunteer to role-play a beginning friendship," I suggested.

He took several shallow breaths and picked a woman named Amy. They sat across from each other, with John deciding the amount of space he needed between them. With my left palm still on his back, I lightly touched his chest with my right fingertips. His body was tight and tense. I asked him to move a little closer and look at Amy.

He slid his chair toward her, then stopped abruptly. "She's a blur. I don't see her at all!" he exclaimed.

"All right," I said, "you don't have to look at her. Close your eyes and go inside your body. Notice how you feel." (Out of politeness, people often continue to look at each other when energetically they have departed.)

He closed his eyes, calmed down and began to breathe more deeply. "Tell me when you'd like to make eye contact with Amy," I said, moving my hands to lightly cradle his right shoulder. He opened his eyes and looked at her, then maintained contact for a little while before glazing over.

"Okay, the moment you're uncomfortable close your eyes and travel inward."

He repeated this exercise several times, and I asked him to locate where his anxiety lived in his body and to touch it. He put his right hand on his stomach, his emotional center, and his left hand over his heart, the center of love. I asked him how he felt. He hung his head down and whispered, "Shame."

Whatever had happened to John as a child produced a deep sense of shame in John the adult. I asked him to go back to that childhood. He recalled being blamed and taunted repeatedly at school. He was forced to stand in front of the class many times and admit that he was bad. Little John internalized the messages from his teacher and classmates: "You're to blame. You're a liar. You should be ashamed. It's your fault."

To survive as an adult he had shut people out rather than being shut out by them, as he had claimed. His body softened when I touched him, as if thanking me for staying by him. He looked into my eyes, nodded and returned his attention to Amy. Now he was able to sit closer and make eye contact. He began to believe his experience: making eye contact would not lead to punishment.

Awareness was an important first step for John. Eventually he learned how to listen to his body cues and to use his newly acquired awareness to change the behavior that had prevented him from having satisfying connections with others.

Inner and Outer Awareness

Awareness of what is around you depends on your needs and focus. It fades in and fades out, a natural flow of changing.

Right now, you are reading this book. Your eyes are focused on this page and they are moving across it. But where are you reading it? In a cafe, on a bus, at home, in a library? The reading becomes the foreground and the noises around you—music, conversation, the rumble of the bus—are background. But your awareness, your focus, can change in a split second.

The following exercise will take you from awareness of your environment to awareness of your thought and then to awareness of your inner feelings.

EXERCISE: AWARENESS CONTINUUM (SOLO)

1. Look around the space you are in. As you notice details, say:

"Right now, I am *aware* of —— " or "Right now, I *observe* —— " (For instance: "Right now I *observe* the blue wallpaper in front of me"; "Right now I am *aware* of a hissing sound.")

2. As you observe the environment around you (the room, light, people, colors, aromas, etc.), begin to include awareness of your thoughts. What are you thinking as you look around?

ıy people cannot observe the environment without opinions or
tations, a deeply entrenched habit.

3. Continue to observe without any judgment:

"Right now, I *observe* blue wallpaper *and* see that some of it is peeling off." (Rather than "I ought to get it fixed," a judgment.)

4. Now purposefully add any thoughts/opinions about the first statement:

"Right now I *observe* peeling blue wallpaper and I *think* that whoever hung it did a terrible job and it has to be fixed."

Practice as many times as you can, bridging the observation with the thought.

5. Continue the awareness/observation sentence, this time adding feelings:

"Right now I *observe* —— " and add "and I *feel* —— " For example, "Right now I *observe* papers all over my desk and I *feel* tired (or excited or angry or relieved, etc.)."

Practice Step 5 as many times as you can, noticing the differences between observations and feelings.

6. Now include all these elements:

"Right now I *observe* peeling blue wallpaper. I think it's time to repaper. I'm *excited* (annoyed, angry, etc.) about changing the pattern."

7. When you are at the "I feel" awareness, close your eyes and travel into your body, allowing yourself to discover where you experience your feelings. For instance: "I feel annoyed" (stomach). "I feel excited" (chest).

EXERCISE:
AWARENESS CONTINUUM (DUET)

This exercise follows a sequence similar to the solo version. However, it will heighten your awareness of how people miscommunicate with each other.

Find a partner who is willing to spend some time with you and participate in this awareness continuum exercise. One of you be A, the other B.

1. Sit facing each other. Agree on the space between you. Close your eyes and take a few deep breaths. Slowly open your eyes and observe each other.

A, begin with the first layer—a simple statement: "I *observe* —— " For instance, "I *observe* you leaning to the right" or "I *observe* that your eyes are twitching." These are observations. At this point, A may—out of sheer habit—add a thought or interpretation: For example, "I *observe* that your eyes are twitching nervously." Of course, B's eyes may *not* be twitching because of nerves. You don't know.

Now it's B's turn to make an observational statement: "I *observe* that your arms are crossed in front of your chest."

A, now respond with another observation of B. Go back and forth like two Ping-Pong players. After several rounds, you may think there is no more to observe. Hang in, though. You may discover more subtle details about each other.

2. Now we'll add the second layer to the observation—an explicit interpretation. (A: "Right now I *observe* that you're frowning. I *think* you're angry.") This layer will help you become aware of the need to take responsibility for making interpretations about others. By including "I" in both the observation and the interpretation phrases, you're clearly taking responsibility for both. (How many times have you heard people say, "You look angry," without knowing how you feel? They are not taking responsibility for their opinion.)

A and B, take turns. Practice this step as many times as you can (as in the Ping-Pong game), paying attention to how you are expressing yourself.

3. You're now ready for the third layer—adding your feelings. ("Right now I *observe* that you're frowning, I *think* that you're angry and I *feel* nervous.") Once again, you are consciously using language to express responsibility for what you observe, what you think and how you feel.

A and B, take turns again. Practice including all three components. When you are ready to stop, take a deep breath and pause. You may want to add one more step.

4. Close your eyes and travel inside your body. Sense where you physically experience your feelings. Gently touch that place and wait for a few moments. Acknowledge your awareness and ask the area if it has any wisdom to share with you. . . . Now slowly open your eyes and share anything you wish with your partner.

Questions to Guide You:

- How did you feel during the observational layer?
- Did your feelings change when your partner added an interpretation?
- Were any of the layers easier to express and hear?
- Were any of the layers more difficult to express and hear?
- Did this awareness continuum exercise help you know more about yourself and your partner?

The ability to distinguish between awareness of what you're seeing, what you're thinking and what you're feeling is the first step toward change.

Who Am I?

The basic question "Who am I?" has challenged philosophers since the beginning of human history. Finding your essence isn't easy, but by shedding some of the unconscious, often crippling weight you are carrying, you can come closer to it.

Much of the personality that seems uniquely yours is in truth the product of the values and behaviors of others inculcated in you long before you were aware of its happening. But to understand them is to understand the fundamental you.

"Who am I?" you must ask. "And what are the roles imposed on me by parents, teachers, society? Why am I in this family? Why was I born?"

Ruth: The Roles We Play

At one workshop, a robust woman in her seventies—unkempt, pale and wearing no makeup—virtually flew onto the table before I had even finished my request for a volunteer. She had never before been in therapy or support groups, let alone attended a Rubenfeld Synergy session. Now she lay on the table, adjusting herself in several different positions, none of which seemed comfortable. Finally, she became still, and I touched her head ever so lightly, discovering a delicate skull connected to a rigid upper neck. When I slid my hands under her tight upper back—behind her heart chakra—she began to cry. I asked her what was happening.

"Don't ask," she responded. "My husband Morty died a few months ago."

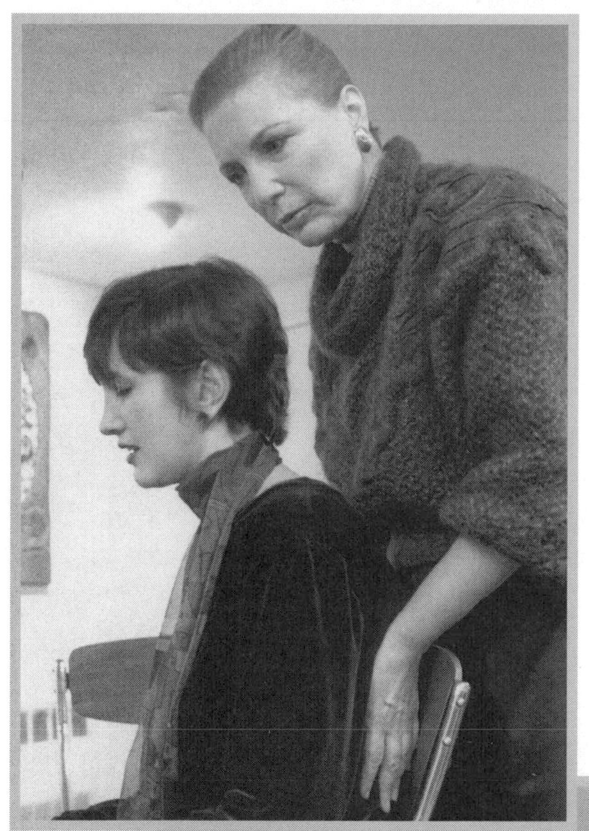

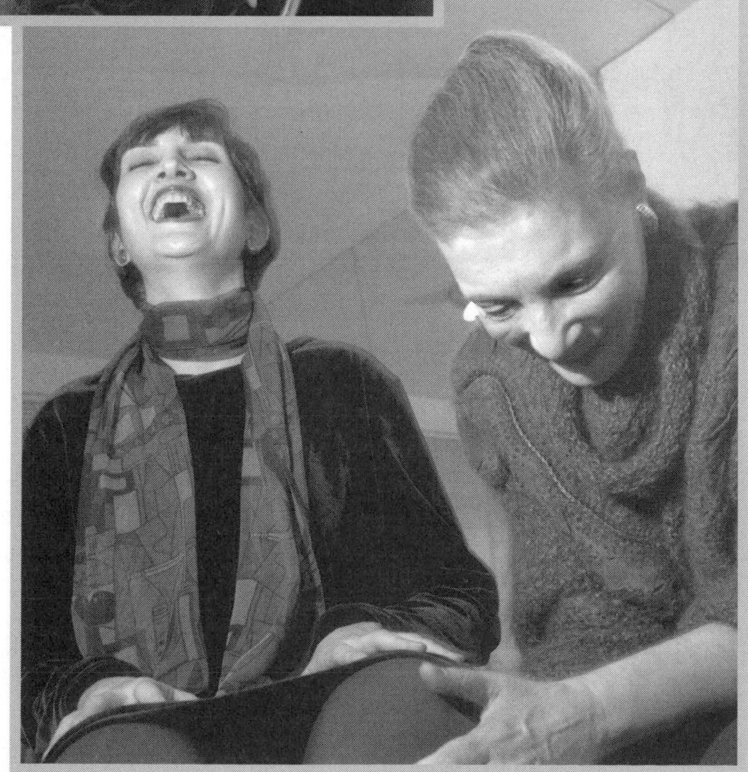

She closed her eyes. "I was such a terrible wife! If only I had cooked him the proper meals, taken better care of him and watched him more carefully—he'd still be alive today."

She proceeded to chant a long list of wifely "shoulds." Her chest was tightening; very little energy was moving into her upper body.

"Did you ever share this opinion of yourself with your husband?" I asked.

"Never. I couldn't. It was my duty to take care of him, and I failed." A new cascade of tears. "I'm a terrible wife!"

"Maybe it's about time you talked to your husband and told him how you feel."

Her eyes opened in surprise. "What? Is he here?"

"No," I said quietly, squelching a smile. "Use your imagination and pretend you can talk to your husband."

After a few moments, she began. "Darling Morty, I've been a bad wife to you all these years. I'm so sorry. I could have kept you alive!" She stopped, overcome with sobs.

"Now take your time. Imagine he is answering you. What is he saying?"

There was a long pause. My hands felt her body soften, more energy moving through and muscles relaxing. Slowly, a smile began to emerge.

"What does he say?" I asked again. Now I was curious.

The smile became radiant as she imagined what her husband would say. " 'Ruthie, my darling. You were the most wonderful wife in the world!' "

Obviously, she yearned to hear this. It was the positive part of herself who was speaking through him. He wasn't negating her role, but confirming she had been a good wife. Her back softened even more and she took several deep breaths. Suddenly, she opened her eyes, looked at me and sat up without any warning.

"Would you repeat what Morty said to you?" I asked, returning her smile.

"Yes . . . 'You were the best wife in the world!' "

"Say it again, only this time change the words 'You were' to 'I am.' "

She was puzzled at first, then she got it. Sitting taller, she said, "I'm a wonderful wife!" Satisfied, she slid off the table and walked out of the room.

We were dumbfounded. Where did she go? As we began to process her session—low self-esteem, guilty feelings, shoulds upon shoulds upon shoulds, unfulfilled wishes—she came back in. She had combed her hair and applied lipstick and rouge.

"Hi," I said cheerfully. "Look at the group. What do you see?"

The answer came quickly. "Warmth and generosity."

"Walk over to anyone in the group and repeat your key sentence."

She picked a woman to approach. "I'm a wonderful wife." Then she moved on, addressing others. By the third person, she changed her phrase to "I'm a wonderful woman." By the sixth, she changed it again: "I'm a wonderful *human being.*"

We all cheered.

Ruth moved from the role of bad wife to a wonderful human being. She discovered her essence.

Different Roles

It often enhances our psychological life to eliminate some of the roles we fall into. We may have outgrown some of them (the "little boy" or "little girl" role seems outmoded in our thirties). Some that we enjoyed will be changed by circumstances ("I loved being a wife, but now I'm a widow"). Some are constricting ("If I go on being a grieving widow, how can I enjoy the rest of my life?"). The point is not to "lose" all the roles we play—some we enjoy, some are important defenses—but to understand that these *are* roles, and thus be able to evaluate them in order to change or eliminate them as we wish.

How many roles are you aware of in yourself? Mine have been daughter, sister, musician, teacher, originator, lover, administrator, conductor, woman, mentor, writer, Jew, traveler, etc. I've made lists of them over the years and continue to ask myself which roles I can eliminate and still be me.

EXERCISE: YOUR OWN LIST (SOLO)

Begin your own list of roles. Write them on a large sheet of paper. Ask yourself what various roles you play in relationship to others.

Take a Magic Marker and cross out all the ones you could live without. Ask yourself, "If I eliminated this role, would I still be me?"

Rewrite the ones you have not crossed out and create a second list. Review this second round and cross out more, this time asking, "Who am I without these roles?" Now look at what remains, and create a third list. Continue to eliminate roles until you are left with those that truly describe your essence, the ones you couldn't live without.

Emotions

Beyond understanding your essential roles, you must also recognize and understand the emotions that underlie them in order to have the capacity to change.

Emotions are energy in motion (E-motion). Infants continually organize themselves to survive, and the need for survival is what guides them to develop contact with others and the environment. (I need food. Will my cry or shout be met?) In adults, emotions are usually energetic responses that mobilize us toward action and are used to avoid or to contact others. Emotions have been named joy, anger, fear, love, etc. When Synergists ask "How are you feeling?" they are inquiring about the emotional states of their clients, not their physical ones, and expect descriptions of emotions in the response. (I am angry, sad, joyful, afraid, etc.) "*What* are you experiencing now?" is another question we ask of clients. "Sadness" might be the reply. "*How* are you experiencing sadness now?" we ask. "*Where* are you experiencing sadness now?" "In my chest," the answer might be.

Notice that we don't ask "*Why* are you experiencing sadness now?" because "why" takes you to the thinking brain, sending it on a search mission for an explanation. Clients often go to "why" first; it is the Synergist's job to redirect them. This is not to say that explanations might not be helpful at another time. However, they are not helpful when you are in the midst of an emotional state. Asking someone "why" they are feeling sad in the middle of an outburst of tears is really saying, "Stop crying." Once the crying has dissipated naturally and there seems to be a closing phase (closure), a "why" may be more appropriate.

Asking clients to change "I *feel* sad" to "I *am* sad" seems to intensify an buildup (crescendo) which moves quickly through to active expres- might want to try this yourself—change "I *feel* happy" to "I *am* nd notice what you experience.

Emotions are valid no matter how they are expressed. A musical phrase sung softly can be as powerful as one sung loudly. One person's single tear can be as moving and authentic as another's cascade of sobs.

Cycle of Emotions

Emotional energy moves through time at its own rhythm and passes through a cycle.

- Awareness of emotional energy.
- Sensing and feeling of the emotion.
- Buildup of emotional energy.
- Highest intensity of emotional energy.
- Shifting of emotional energy into action or nonaction.
- Decreasing of emotional energetic intensity.
- Completion (closure) in both body and psyche.

Becoming Aware of Resistances/ Defenses to Change

Sometimes, however, even if we recognize our emotions, we might want to flee them. They may be too painful—or, paradoxically, too joyful—and their intensity too frightening. We resist them, defend against them, refuse to change as they dictate.

Traditionally, resistances and defenses are defined as stubborn barriers that prevent people from becoming their authentic selves.

In Rubenfeld Synergy, resistances/defenses are accepted as part of people's identity, not some "stubborn barrier" to be removed. The defense formation is ongoing; it starts in infancy and continues throughout our adult life. We've seen that babies make contact with others in order to survive. When contact is disrupted or not forthcoming, they develop ways to cope which may become dysfunctional. Many defenses are habitual and unconscious; we are not aware of them because they were developed so young.

Adults may behave and function as if the threat to survival they felt during infancy still exists when in reality the threat has disappeared. Or they may be defending themselves against threats from their adolescent years—or even the day before yesterday. We will be looking at several resistances/defenses that people use to prevent contacting others, themselves and their environment in an authentic way.

Defenses exist in our minds, bodies and psyches. By creating safety and trust and accepting people where they are and respecting their resistances, Synergists lessen the power struggle between themselves and their clients. A tight body area may signal a client's resistance to change even after the rational mind understands the reasons for the tightness and why the area became tight in the first place. The listening hand acknowledges the resistance and does not force, confront, fix or overpower the client's physical way of being. The use of touch gives us important feedback about physical and energetic defenses while communicating patience.

I teach that one of the most important techniques we have is *waiting.* During these waiting times (like silences in music), my hands sense whether the area is too tight to change at this moment, and I may move to another place that is less tight and more available to contacting, connecting, relaxing and letting go.

Using touch to encounter resistances presents unique challenges and opportunities. Although the listening touch begins from the outside in (the skin or clothes), it serves to heighten people's awareness from the inside out, focusing their attention on a specific contacted area. Their bodies experience the toucher's creating a safe physical and psychic space where they can dialogue without limiting thoughts and feelings that have previously been out of their awareness.

At a workshop not long ago, Jerry complained about a pain in his shoulder. I touched it gently and introduced a simple verbal dialogue:

"If you could talk to your shoulder, what would you say?"

"I wish you didn't hurt so much!"

"Give your shoulder a voice and let it answer you."

" 'I'm tired of your heaviness—shouldering all the burdens in the company. I want some fun and pleasure too.' "

His tight, heavy shoulder is a wonderful metaphor for Jerry's life and embodies his resistance to fun and pleasure. I don't know his story yet, or the reasons for the resistance. I *do* know that his shoulder spoke the truth and that we can now go on to explore the defenses/resistances that Jerry uses unconsciously to avoid fun and pleasure.

There are many forces, often received and acted out unconsciously, that demand change, even though we don't know it.

Introjection

From the time you were born you were fed psychic food. Without chewing, tasting and assimilating, you swallowed it whole. Children are like sponges. They absorb ideas, behaviors, actions, values and postures and express them as if they were their own. Rachel, a six-year-old neighbor, came up to me the other day and cheerfully announced, "I'm a Republican!" This defense/resistance is called *introjection.* It begins at birth and continues throughout life.

Lois, a close friend, vowed to never criticize her children the way her mother criticized her. "I was yelling awful negative and critical words at my kids when I suddenly stopped and heard myself talking just like my mother."

Negative introjects are powerful and create the critical voices in our minds and bodies. They shape what we do and how we behave, binding us more securely than chains.

Much of our unconscious behavior and coping as adults derives from childhood defenses we devised to protect ourselves from pain, hurt and injury. Our parents continue the legacy and tradition of *their* parents, adopting unconscious dysfunctional behaviors and passing them on to the next generation.

"It's too risky, I'll get hurt," we think before starting a business venture, forgetting how our timid parents cautioned safety all the time, even when we climbed on the jungle gym. "S/he'll never be attracted to me. I'm not handsome/pretty enough," we tell ourselves, forgetting how our parents would compare us unfavorably with our sibling.

Take a moment and return to a memory of an introject. What did your parents (or teacher) say or do? Where do you feel this in your body? Awareness of such introjects will help you understand your behavior.

Retroflection

Doing for yourself what you wish others would do for you is called *retroflection.* When children express intense emotions, these energies move toward the source (parents, teachers, peers). If they are unacceptable to either the child or the source, the emotional energies retreat inward, and so instead of being angry with others, for example, you are angry with yourself. My mother was often furious with my sister and me. Instead of expressing her anger, she gave us the "silent treatment" and became depressed, thus hurting herself.

Steve's mother was left by her father when she was a toddler, and she invested her husband—Steve's father—with all the qualities she wanted to believe her father possessed. She continually indicated verbally and non-verbally that Steve was inferior to his father—her husband—in every way. Whatever he did was never enough. To please his mother, he (usually unconsciously) imitated his father whenever he could, including his walk and posture—shoulders slumped, chest concave—and ingratiated himself so much he eventually could not locate his uniqueness. Since his father was often on the road and his mother was not available for any support, he quickly learned to do everything for himself and not trust anyone else. (This can be a positive side to retroflection: if your wish to have someone take care of you is unfulfilled, you will usually learn to take care of yourself.) As a child, this retroflection saved him. As an adult, having to control and do everything himself, he was continually harried and never slowed down. Defenses helped him through childhood, but as an adult he was dominated by his earlier behavioral pattern. Not being able to count on others to help support you may result in stress, loneliness, continual worry and an inability to make authentic contact.

Steve struggles to change—*wants* to change—and is succeeding. But sometimes he reverts to the "old" way. It is a continuing process, challenging, exciting, difficult, flawed. And well worth it.

Projection

"You look angry. Are you angry?" This is often said by people who are themselves angry but don't realize it. "Do you still love me?" they ask repeatedly, though they feel no love for the other. They project their feelings, opinions, attitudes and behaviors onto everyone else. They are like slides projecting their own images onto screens, not acknowledging themselves as the source.

Projecting your emotions, feelings and ideas prevents contact between you and others. "They"—the others—are the bad or good ones, not you. *Projection* is debilitating when you don't recognize how *you* truly feel and continually ascribe your feelings to other people or to groups. But the more you recognize that your own feelings are the basis for the feelings you ascribe to others, the greater your chance of dealing with your inner feelings without distortion.

Confluence: The Urge to Merge

Sometimes, in a psychological need for acceptance, people adapt the behavior, thoughts and feelings they perceive in others, even though these behaviors, thoughts and feelings might not be their own.

For example, Matthew, the son of parents who taught him that being a nonconformist was "dangerous" and that nobody would like him, was interested in classical music. But he was afraid to behave differently from his friends—he would not be accepted by them, he thought—and so listened to rock and told himself he liked it. Growing up, he lived in two worlds. His inner world housed his true feelings, dreams and yearnings. In his outward world, he merged with others. His body mirrored this behavior. He began to sweat anytime we approached the subject of expressing what *he* wanted instead of mimicking others. After many sessions, as his body learned to soften, he began to experience some relaxation and pleasure. He could express his opinions and behave authentically without anxiety or fear. By contacting his true self, he was able to contact others. He moved from a rigid life pattern to a more flexible one, slowly going through the stages of the change process.

People who are *confluent* change their opinions, behavior and "colors" like chameleons. They do not distinguish their own thoughts, attitudes and feelings from those of others. You hardly ever get to know the person inside, because confluent people do not know themselves.

Susan walks into my office, sits down and exclaims immediately, "We have so much in common." She's a therapist like me, uses bodywork, loves my furniture, paintings and color scheme. She announces proudly that she's able to "merge" with her clients and can "feel their pain." Their discomforts become her discomforts, their problems her problems. Susan believes the similarities between herself and her clients are uncanny. She does not experience herself as different from others, but the same. She is a confluent person.

When confluent people meet you they are saying, "Hello. How am I?" They wait for your opinion before expressing theirs. Who in your life is confluent? Do you really know how they feel or think?

Confluence can also be a creative force. Great actors or mimes are confluent as they become the movement, furniture, posture and soul of the people they are portraying. In the words of Fritz Perls, "As in Zen, to paint a leaf you must become it."

Malevolent Transforming: From Gold to Garbage

People who continually hear compliments, nourishing statements and genuine concern as negative and attacking are *malevolent transformers.* Between the time a "golden" statement leaves your lips and reaches their ears, they have turned what you've said into "garbage." They transform and twist positive and well-intentioned offerings into pain. They ward off the possibility of receiving love from others. Sometimes you may find it best not to continue saying positive and supporting things because they just add more fuel to their painful fire. This defense is shaped like a bowl that has no bottom.

"You look beautiful today," I say.

"Yes, but you can't mean it. I'm exhausted today. You should have seen me yesterday when I didn't have such dark rings under my eyes."

"Yes, but" is the signal for what is to come. In my family, you never complimented the cook (my mother)! The response was invariably a list of things that went wrong, one disappointment piled on top of another.

Deflection

A little girl knows that when her mom and dad have an argument, she had better not let her mom brush her hair. When people are not able to express their feelings directly, they may place them inappropriately onto someone else who has nothing to do with the original problem. This is *deflection.*

I was riding home from the airport and the cabdriver was unusually aggressive and nasty, swearing and honking his horn continually at the other drivers. I asked him to stop. He did, but he also turned to me, enraged, and muttered a curse.

Direct confrontation was my reaction. "Look," I said. "I haven't done anything to deserve this kind of behavior. Maybe you've had a rough day, but I didn't cause it. Unless you stop your hostility, I'm getting out." Surprised and somewhat amused, he apologized and drove more calmly the rest of the way.

Using Defenses Positively

Return to the awareness continuum exercises (pages 95-97) and look at each step with a new understanding of your defenses. "I *observe* that you are looking at me AND I *think* that you are angry with me." The *think* response may be a projection.

Stop, close your eyes and ask yourself, "Am I angry?" "Am I sad?"

Introjection, retroflection, projection, deflection, etc. can be beneficial rather than destructive. The person introjected with love will project love; the neglected child may be the self-reliant adult. The point is that by becoming aware of these defenses in yourself, you will be able to change the negative into the positive and use the concomitant energy for a more vital life.

8.

Seven Steps to Change

Classical, jazz and popular composers have come to appreciate and use a form called "Theme and Variations." In the seven steps of change, there is a main theme—change itself—with different variations, all having a relationship to each other.

Inner change is a cyclical process, not a linear one, which builds from each step. You may find yourself spending more time in one step than in another—and then going back to the beginning of the cycle before you can proceed again. Repetition is often necessary in any or all of these steps.

Step 1. Awareness

As we've seen, awareness is the first step in the change process. Without it, no change is possible. You are trapped in the status quo. When you sense, recognize and become aware that your life and behavior are not satisfying, it is time for change. When all is not well within your body/mind, family and environment, it is time for change. When you realize that your life is a treadmill and you are living in the same status quo, it is time for change. Sometimes awareness alone can spark changes in behavior and lifestyle, but usually we need more steps to effect permanent, useful change.

Willingness is a vital partner of awareness, for your will plays an important role in change. To change a habitual way of life demands commitment, a willingness to experiment, to recognize and use your inner resources. There will also be times when you may need outside help—a close family member, a friend or a professional—to guide and bolster you.

Step 2. The Challenge of Nonhabitual Events

Outside sources, such as serious illness, separations, the birth or death of a family member or close friend, changing environments and career shifts (to name just a few), are another important input that will disrupt your inner and outer status quo and balance. At such junctures, your habitual ways may not want to change; after all, they've been with you for a long time and served some important functions. Still, they were learned, and can be unlearned when they're no longer needed.

Step 3. Confusion and Chaos

I don't remember ever getting a gold star at school for being confused, even though I often was. In fact, confusion is essential for change—indeed, for all learning and growth. The word "confusion" has two parts: "con," meaning "with and opposed to"; and "fusion," meaning "joining." Confusion becomes an important step in the change process, for you must jostle your thoughts, body experiences and emotions in order to fuse them differently to effect change. The habitual tensions of our lives are organized early on in a particular way. In order to dislodge and change them, one must go through a time of dis-organization—chaos. Many life habits come from early unconscious messages. One powerful way to change those messages is to distract and occupy the attention of the conscious, rational brain. In this state, you have the opportunity to experience different emotions and discover another way of being. You must be willing to leave your familiar territory and navigate through chaotic and confusing waters to the new land.

Step 4. The Fertile Void

This is a time when your old ways do not work and you don't have new ways of coping. It's in this "in-between" space, moving from one state to another, that you need encouragement and support. This "I don't know where I am" state creates anxiety for most people, because it is unknown territory, far from the familiar. Since you have not yet practiced or integrated new behaviors, fears, doubts and anxieties may and probably will emerge. The fertile void is a place of change in which one sometimes feels "stuck." It's tempting to push through this stage quickly, to deny the struggles, fears and doubts. By experiencing it fully, you will be able to continue on to the next stage.

Step 5. Practice

Once you have moved through the fertile void, you are ready to use your inner resources and all the skills you have learned and to practice a different way of behaving. This is an important time to have trusted allies with you, because previous defenses may continue to haunt you as you rehearse, face past demons and perhaps get confused again. If you go inside, communicate with your inner self and renew the feelings of safety, you will be able to sustain contact with the new you. You may have to practice this different and new sequence several times before you feel confident of your new discoveries.

Step 6. Integration

After you become aware and learn to recognize the various roles and behaviors you have internalized, you can experience relating without using these dysfunctional defenses. Now you are ready to integrate and assimilate new insights and different behaviors. In this integrative phase, your body is a great barometer of the truth, for it experiences and recognizes these changes before the mind does. A sense of wholeness and well-being begins to develop. The fear of not surviving diminishes and you can truly release your old defensive tactics.

Step 7. New State of Being

Once you have passed through all the previous steps (and repeated some of them), your new way of being will soon feel more comfortable and familiar—

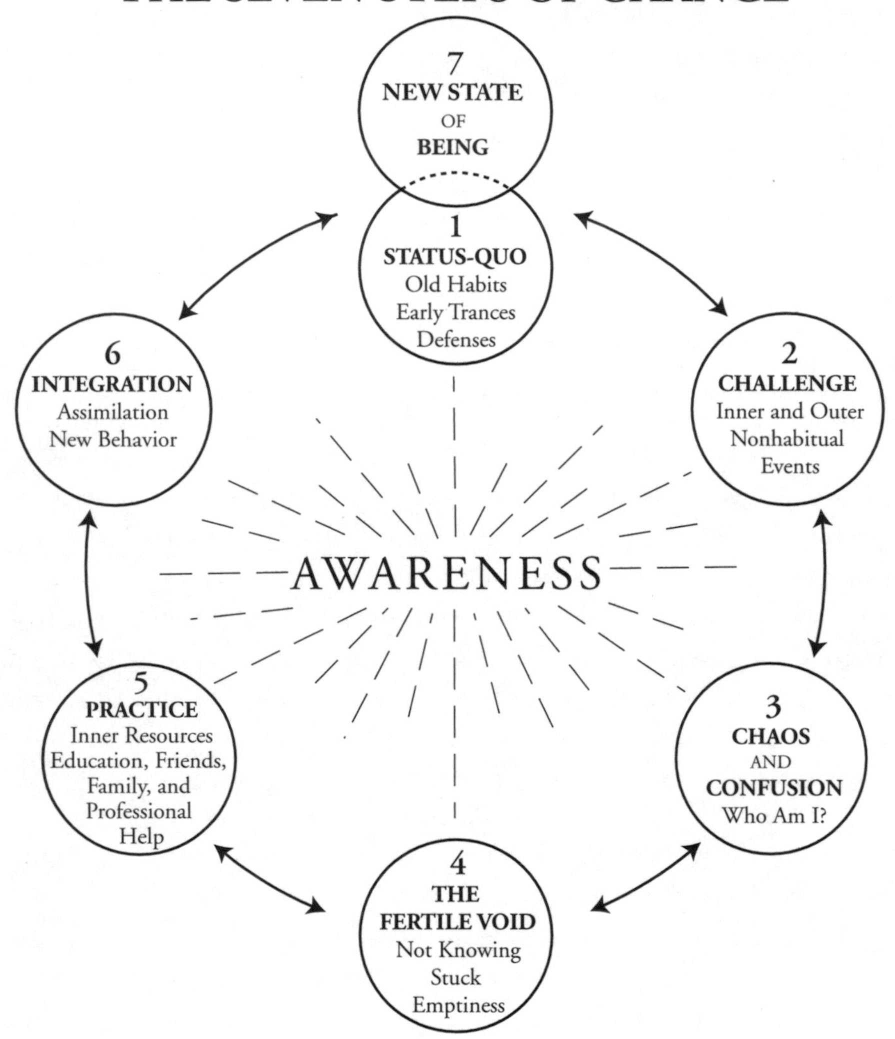

perhaps even exciting. I have seen many people be willing to go through these steps because of their strong resolve to change. Your change may cause a ripple effect on your family and friends. They may fight your changes; they may feel abandoned. They may or may not change dramatically in response. Each Rubenfeld Synergy session usually includes some of these steps, and many times Synergists will guide their clients through all of them.

The use of talk and listening touch supports clients in their effort to deeply change. Also, touch informs the Synergist about the readiness of a client's body to accept and integrate changes.

As you rebalance your life, your changes may improve your marriage, create more contact and intimacy with your friends, allow you to be more forceful at your job or even enter a new career. Without question, the changes will influence the way you work, live and love.

Taking Responsibility

A way *not* to change is to blame other factors: the environment, the school, the government, your parents, spouse, children, friends and so on. When was the last time you heard people blame a force outside themselves for their predicament? When was the last time you did it yourself? "This feeling just came over me" is a common statement clients make, as if the "feeling" were initiated by other forces and not the person feeling them. Clients often say, "If only I could change my parents (my partner, my children) I would live a happier life." After some time, it dawns on them that the only person one can really change is oneself.

Many of our defenses have contributed to ways of not fully taking responsibility for who we are and what we want. And your body may become tight and unyielding when it takes on the heavy burden of thinking that everyone else "should" change.

Once you are aware of your blaming habit and take responsibility for it, you have a chance to change. A willingness to "own" your actions and look within are vital to the change process. (On the one hand, I work to increase "ownership" and responsibility, but on the other I often find myself helping clients sort out what they are *not* responsible for. Because children tend to feel responsible for many bad things that happen to them, learning when others, not you, are wrong is just as important as acknowledging when you are wrong.)

This notion of responsibility, however, can be simplistic when taken to the extreme. Subscribing to a popular self-help saying like "You always cause your own illness," for example, may engender guilt, not healing.

Vicki: "It's Not My Fault. I'm Just a Baby."

Many years ago, the procedure for treating premature babies was to put them in an oxygen-filled incubator. It was cold in the incubators; mothers were not allowed to touch their babies. And the oxygen could cause damage, as it did in Vicki's case. When she emerged, she couldn't focus her eyes and had severe difficulty seeing. She was proclaimed legally blind and had worn thick lenses since infancy.

The adult Vicki was brave. She received a master's in social work and completed the Rubenfeld Synergy training. One day, she called me, sobbing.

"I can't see, and it's my fault," she wailed. "What am I going to do?"

I invited her to come to the center so we could work together. There her story emerged. She had just completed a five-day intensive seminar based on the proclaimed belief: Everything That Happens to You Is Your Doing and Your Responsibility.

"What is it that you don't want to see?" her facilitator asked Vicki. "How did you make yourself blind?"

Hysterical blindness has been well documented, but this was certainly not the case with Vicki. After some wonderful and hopeful changes in her life, she was sliding back into a feeling of helplessness, like the baby in the incubator. Her body trembled when I touched her. She was very cold. As a baby she had wanted to be touched and held; no one had come to her.

She began to move her arms around. "I want these tubes out. They hurt!"

I pulled out the imaginary needles and Vicki felt some relief. She began to relax.

"It's not my fault," she cried. She was still shivering, and I covered her with blankets. Finally, warmed by them and by my touch, she opened her eyes, peered into my face, then slowly looked around the room. "Colors look different now," she said.

I asked her to repeat "It's not my fault" and add "I was just a tiny baby and I did not cause my blindness."

Many survivors of abuse blame themselves for their physical and emotional condition. Vicki did just that—blamed herself for her blindness. When she finally recognized that baby Vicki could not have caused her condition, she stopped blaming herself. She became more accepting of her situation and gave herself "permission" to have more pleasure in her life. These changes expressed themselves dramatically through her eye condition and her specialist was eventually able to fit her with contact lenses which opened up more of the visual world to her.

Mary: "I Just Want to Hear Positive Feedback."

How powerful the process from self-blame to self-knowledge was reinforced for me at a recent workshop. Mary, a stylish woman in her early forties, complained that it was becoming more and more difficult to interact with a mentor she had relied upon in business. I had her sit facing an empty chair and told her to imagine she was facing him.

"What would you like to say to him?" I asked, lightly touching her lower back, which was arched.

"You're always telling me I'm not doing enough, that I need to improve, that I'm not measuring up," she said in a whiny, high-pitched voice.

I suspected some early negative introjects in her life. I moved my hand to touch her upper back. It was stiff and tight. Her legs were crossed and she held her hands on her lap, fingers tightly interlaced. She kept her head down and spoke in a whisper.

I moved to her right side and placed my hands lightly on her shoulder. "What else would you like to tell him?"

"I've outgrown you. I'm not going to keep being with you."

Her back rounded, she made herself small and her voice was very high-pitched. She looked like a little girl, and flinched as if bracing for a blow of some kind. I went to her left side and placed my left hand in front and my right hand in back of her left shoulder.

"Was there anyone in your childhood who spoke to you like that?"

"My mother!" she said quickly, still in a childlike voice. "There was nothing I could do that was right." She paused, remembering. "I got good grades in school—they weren't good enough. I excelled in sports—I was a klutz. When I was twelve, she put me in a terrible school and made me go to work in the afternoons."

"That's right," I said, hearing fury in her tone. "You can get angry at her." I made some sweeping movements upward, an inch or so from her body, assisting her "angry" energy to move.

She turned her head and glared at me. "I've worked on this before. Getting angry doesn't help. It has no bottom. I can't seem to get beyond it. My mother may have been critical of me, but now I'm supercritical of myself." Her body was shaking and her head hung down again. I touched the back of her head.

"What about your colleagues?"

"They say nice things sometimes. But I don't believe them. I don't even *hear* them."

She suddenly looked up at me, obviously wanting some kind of rational explanation. I moved my left hand away from her head and placed it on her upper back. "You've swallowed your critical mother, and she lives in you. Not only in your head, but in your body. She's the critical voice coming out of your mouth."

Her body began to soften as she took this in. Her eyes held a mixture of confusion and fear. "What can I do?"

"This began when you were a child and you were in an undeveloped state and took the negative voices in," I said softly. "You will have to interrupt this loud critical inner voice and change this negative way of looking at yourself. You *can* rewrite the script of the past, because you are the writer and the star," I said enthusiastically.

Her body began to straighten and become flexible, welcoming. "Remember, by rescripting what happened in the past you can change yourself now," I explained. I stood behind her, touching the sides of her head. "Every cell in your body is ready."

I gently moved her head to the left and to the right. "On which side do you hear your mother's critical voice?"

"The left," she said instantly.

"The moment you hear her critical voice in you, turn your whole body to the left—and deal with your mother's negative side."

She swiveled around to the left, facing her imaginary mother. I took my hands away from her head and continued to stand in the space by her right side. "Can you think of anything supportive she said to you?" (There's always a shred of something positive to cling to.)

"No. Nothing!" she said adamantly, returning her body back to the center. I touched her right shoulder again.

I laughed. "There must have been *something.* Otherwise you'd be in a corner sucking your thumb."

A small smile crept onto her face. "She always told me I was a good worker," Mary said slowly, turning squarely to the right to look at me. "That I accomplished a lot."

She sighed, and I could feel her upper back and chest growing softer still. Suddenly she frowned, and her body immediately tightened. I knew she was hearing her negative critical voice. I quickly stood behind her again and moved her upper body, head and neck to the *left.* "Oh my God," she gasped. "I didn't know it could happen so fast. And you could tell?"

"Of course. The moment you went from the positive statement 'I accomplished a lot' to hearing a critical voice, your whole body changed. You can interrupt that voice now!"

"Leave me alone!" she shouted to her negative imaginary mother. "Go away!"

"Great," I said, returning to her right side. "The minute you experience that negative voice turn your whole body to the left side and confront it. Tell her to go away or anything else you want."

Still facing left, she repeated, "Leave me alone," but this time in a far less angry tone. She sat up, straightening her back and head, no longer a hunched woman shielding herself from harm. I touched her right shoulder. She turned to me.

"You're a wonderful athlete," I told her in a soft, loving voice, remembering that she excelled in sports. Role-playing her protective, good mother, I added, "I want to help you combine your good grades and your sports." I leaned toward her right ear, still touching her right shoulder, and repeated these phrases several times. "Where are you experiencing these words?" I then asked.

"In my gut and heart." But suddenly her critical voice interceded. With a surprised cry she swung around to her left. "Oh my God," she said. "She's done it again." Her expression became resolute. "You've been in my body for forty-five years. That's enough. Go somewhere else!"

"Not me!" a man in the group exclaimed spontaneously. Everyone laughed, Mary included. She returned to her center position, sitting tall and squarely in her chair. I rested my hands on my lap.

"Now imagine that your mentor/teacher is sitting in that empty chair and tell him what *you* want to hear," I instructed.

She looked at the chair. "I want to hear positive feedback from you. No more criticism. No more blaming me." Her eyes were shining.

I asked for a volunteer to role-play her mentor and sit to her right. "Tell your mentor what you want to hear."

" 'Mary, you're so bright and intelligent. You've resolved the problem in such a creative way,' " she said.

He repeated her phrases exactly. Standing behind her, I put my listening hands on her shoulders. They were relaxed. Her body was totally taking in the positive comments. Her emotions, mind and spirit were integrating them. She turned again to her left and without instructions from me told her critical voice to go away while spontaneously making a powerful downward movement with her left hand—cutting off her mother, cutting off pain. Her nonverbal gesture was a wonderful way to stop these negative thoughts. I told her to use it anytime she heard that voice. She laughed and said it would look peculiar during business meetings.

"Do it under the table," I said, smiling and sliding my hands off her shoulders. "You've got to counterattack the negative as soon as you hear it."

She made the slicing motion again, turning to look at me. "My body feels so relaxed. In all these years of anger, I never felt like this. I feel strong enough to really tell my mother to stop criticizing me all the time and start praising me for my work."

I nodded. Whatever introjections she had been fed by her mother—facial expressions, physical reactions, degrading words—had obviously taken up her psychic space and inhabited her body, stifling pleasure and psychological nourishment.

The shift from negative to positive had to be made at the cellular level as well as in her rational mind. Mary, I am sure, will continue to use these physical cues, continue to cut off her mother's cruel words, and stop being so critical of herself.

Over the years, what I did with Mary has become an important technique: assigning different messages to different sides of the body.

From Negative to Positive

Negative critical inner voices are introjected at an early age—looks, attitudes, nonresponses, sarcastic tones, body language and *words.* It's true that "sticks and stones will break your bones," but hostile words *do* harm us as well, especially at ages when we're most vulnerable: early childhood, our preteen years, adolescence. "You're stupid." "You'll never make it." "You're no good and you shouldn't even be here." "No matter how hard you try, you'll never succeed." "You're not as pretty as your sister." "Why don't you play piano like your cousin?" "You don't measure up to your father." "I won't love you unless you join the family business." These are only a small sampling of what I've heard at workshops. The phrases are part of old, early scripts enacted by various family members, teachers and friends.

In the following exercise, you will be engaged in a transformative process, using an actual rewriting of your inner scripts. I've created the exercise for you alone and for you and a partner.

EXERCISE: TURNING FROM NEGATIVE TO POSITIVE (SOLO)

Find a quiet space where you will not be interrupted for about a half hour to an hour. Have paper and writing tools ready. If you keep a journal, have it nearby.

1. Sit down comfortably, close your eyes and take a few deep breaths.

2. Travel back in time to a scene when someone said negative words to you, words that hurt. Notice the person or people who spoke, and what they looked like as they said those words. Allow yourself to feel any emotions, giving yourself time to locate them in your body.

3. Repeat the negative words aloud or to yourself and become aware of where they are in your body. Where do these words live? Where are you carrying them?

4. Now we'll use the technique of two different sides. Each time you say or think the negative words, move your body in a definite and clear direction (which will remain the same). Imaging the person or situation, say anything you wish to yourself or aloud and notice how you feel. When you have a sense of completion, return to your center. (In Brazil, the healers shake out negative and harmful messages. Try this if you wish. Then rest.)

5. This time ask yourself what positive words you really wanted to hear. Turn your body in the opposite direction, signifying a shift to the positive side. Say your positive words aloud or to yourself several times and notice how you feel. Where does your body experience these positive words? Take a few breaths and repeat your positive words again. Notice if there are any changes in your body.

6. If the negative words return, turn your body back to the negative side and express your determination to oppose them.

7. Now move yourself to the middle and rest. Slowly open your eyes. Reach for your paper or your journal and fold the sheet in half, creating two columns. Write your negative introject in one column, then say it aloud. Pause. Are you aware of anything left over from the old negative script? If so, put the paper down for a moment, turn to the negative side and have a dialogue with it. When you have finished, take a deep breath and return to the middle. Pick up your paper and write the positive words in the second column. As you write, speak them as many times as you wish, noticing how your body responds. Allow yourself time to express your emotions.

8. Close your eyes and return to a quiet sitting position. Image people saying positive words to you. Whisper them, shout them, say them softly.

9. Open your eyes and write anything else you wish. When you finish, close your eyes again and rest. After some time, notice how your body feels, open your eyes, look around, get up and move.

Transforming Negative Messages to Positive Ones

(*From Workshop Participants*)

Negative Introject	Positive Rescript
• You're stupid	• You're learning quickly
• You'll never make it	• You can do anything you want
• You're no good and you shouldn't be here	• I love you and I'm glad you're here
• No matter how hard you try, you'll never succeed	• You'll succeed at whatever you decide to do
• You're not as pretty as your sister	• You have your own beauty, and it's unique

- Why don't you play the piano like your cousin?
- You don't measure up to your father
- I won't love you unless you join the family business
- I won't love you unless you do what I want

- I really enjoy your piano-playing. What a treat!
- In many ways, you are better than your father
- I'll love you no matter what job or business you choose
- I'll love you no matter what you do

EXERCISE: TURNING FROM NEGATIVE TO POSITIVE (DUET)

In this version, your partner becomes a powerful ally in rescripting your negative voices. You will have an opportunity to tell your story and hear your positive phrases said aloud to you.

Sit facing each other with one to two feet of space between you. Decide who will be the first storyteller and who will be the first listener. You must agree upon several points in order to create an atmosphere of safety and trust.

The Storyteller: Share anything you wish. If you are uncomfortable at any point you may want to stop for a while before continuing.

The Listener: You are here to listen to your partner without judgment or criticism. Let your intention of compassion and care be your guide while listening. Accept what the storyteller relates. If your partner is emotional, do not interrupt, ask questions or give advice.

1. *Storyteller:* Close your eyes, take a deep breath and travel back in time to a scene in which someone said negative and hurtful words to you. Listen to your body as you reimage the scene and hear the words. Allow yourself to feel any emotions and give yourself time to become aware of them in your body. Turn your body in the direction of the negative voice and say anything you wish to that space.

2. *Storyteller:* Open your eyes and tell your listener the scene, person or people that emerged. Specify who, where and what was said. What was the negative, hurtful phrase that you heard? If you feel any emotions, express them to your listener. Again, notice where you carry the negative phrases in your body.

3. *Listener:* Your role in this part is to listen without judgments, opinions or advice—to just be present with empathy and compassion. Perhaps your partner's scene reminded you of your own story. Take several deep breaths as you continue to listen; you will have a chance to share your story later.

4. When the storyteller is finished and has identified the negative phrase clearly, we will begin the transformation process.

5. *Storyteller:* Close your eyes. Rescript the same scene by changing the negative, hurtful words to supportive, loving and positive words you really would like to have heard. Turn your body toward the side which is positive and repeat the new supportive, loving words several times. Then return your body to the middle space.

6. *Listener:* Move your chair to your partner's positive side. Sit down and reposition yourself so you are facing his/her shoulder. Ask your storyteller to repeat the positive phrase again. Now lean close to your storyteller's ear and say the positive phrase several times.

7. *Storyteller:* Allow your body and mind to hear this positive phrase. If you want to hear it more times, tell your listener to repeat it.

I've seen many people at workshops get upset at first when they hear the loving, positive phrase they asked for. Although they yearned for a parent, for instance, to say these words, they are not used to hearing them and may need more time to believe and integrate them. Sometimes, when I am the listener, I add touch to this section so that the storyteller receives the same positive message through my listening hands.

8. *Listener:* Ask permission to gently touch the storyteller's shoulder while you say the positive phrase. Lean toward the storyteller and gently place one hand in front of and one hand behind the storyteller's shoulder. For example, if the negative, critical phrase was "No matter how hard you try, you'll never succeed!" the storyteller has changed it to "You'll succeed at whatever you decide to do!" Repeat the positive phrase many times while touching the storyteller, asking your hands to send the same positive message.

I have suggested that the storyteller add another phrase to the main one:

"And I'll always love you no matter what you decide to do." This has become a meaningful addition for many people.

Repeat both phrases until the storyteller signals you to stop. Move your hands away slowly. Rearrange your chair so that you are sitting facing your storyteller.

9. *Storyteller:* Allow yourself to remain quiet. After some time, open your eyes and see your listener-partner. Share anything you wish with each other.

Questions to Guide You:

Storyteller:

- How did you feel being listened to in such a caring and focused way?
- Did you confront your negative voices? If so, how and what did you say and feel?
- How did you experience hearing the positive words?
- Was the touch/contact helpful?
- What inner resources have you discovered?
- How will you use them when you want to transform a negative, hurtful message to a positive one?
- How do you experience yourself right now?

Listener:

- How did you feel listening without judging or offering comments or advice?
- What were you feeling, thinking or sensing when you heard the tale of the storyteller?
- What were you aware of when you repeated the positive phrases?
- How did you feel using touch while repeating the positive phrases?
- What did you observe about the reactions of your partner?
- How does your partner appear to you right now?
- How do you experience yourself right now?

After this period of sharing aloud, it is time to do a de-roling process.

The listener role-played the positive teacher, good mother/father or beloved by constantly occupying the positive space, remaining caring and continually repeating only the transformed positive phrases. The process of de-roling is important because it sets listeners free of always being in these roles and helps them return to themselves.

The listener says, "I am no longer your ideal, positive parent (child, boss, spouse or beloved). I am (name), your friend."

The storyteller realizes that the listener is no longer playing this role, but is simply a friend again.

Take a break. You may want to write in your journals, or sketch.

When you complete the entire process, thank each other.

After some time (which you will agree on), begin the duet exercise again, reversing roles.

In private Rubenfeld Synergy sessions, Synergists will often take the role of listener and repeat the positive messages to their clients.

Move at Your Own Pace

As I've said earlier, emotions have energy and loving phrases may be overwhelming. So at first, move at your own pace. Rescript the past scenes in small portions so you can assimilate and integrate the positive, nourishing words that touch your body and soul. Positive feelings are wholesome food for the body and brain, and the more pleasure you can incorporate, the healthier you will feel.

"Thinking" your way to change is most improbable without the participation of your body, your feelings, experience and practice. The exercises in this chapter will help you on your journey of discovery and change within your body and mind.

Still, even with understanding, experience and practice, there will come a time when a family reunion, a business crisis, a fight with a spouse or child sets off familiar alarms. The negative voices raise their ugly sound and there you are, acting in the old habitual ways.

Another tool is required if deep and lasting behavioral change is to occur: the therapeutic trance.

9.

Trance and Altered States of Consciousness

When I first see her, the word that pops into my head is "clean." Her blond hair is brushed and shines in the morning sunlight flooding the windows of the workshop room. She is wearing sandals and I can see that her toenails, like her fingernails, are cut and polished, as is a silver toe ring. She wears a white dress as pristine as a wedding gown, and there is a sparking silver locket attached to a chain around her neck.

Strangely, though, she stands to the side of the room, not joining in the morning exercise and class. She watches the group intently. We have begun the session with exercises, but now the group is dancing to the Brazilian music on my tape recorder. They form a conga line as the beat intensifies, obviously enjoying themselves. The woman is moving her feet in rhythm, but she is alone, in effect the only unconnected person in the room.

Later that morning she lies on the table, arranging herself carefully so that the dress comes far down on her calves and will not wrinkle. She is both eager and recalcitrant as a volunteer. I approach her cautiously, lightly touch her head. At first she flinches at my touch, but then she accepts it. She is already on her way to the first level of a trance state.

Her name is Janet, but when she first introduced herself she told the group that she liked to be called "Sweetie."

"I hear energy in your hands," she tells me.

"Mozart?" I reply, smiling.

"It's electrical."

"I feel your energy too. Every part of you has its own vibrations, especially your toe ring."

She laughs for the first time, relaxing a bit. "It was an earring first, but it belongs on my toe."

Now, I can touch her feet. They're delicate, like her hands. I move up to her left hip and ask her to take a breath and say anything she wants to her hip and pelvis.

"Relax," she says immediately, but my listening hands reveal more tension than before.

"How do you experience your pelvis responding to the word 'relax'?"

Her mouth tightens. "It's closing up."

"Then your pelvis isn't sure what the word 'relax' is yet. Say something else."

"Color purple!" She pauses. "Yes, purple." Her hip joint softens.

Gathering her energy, I move it through her leg and out her foot. "Okay, one side is purple." I cross slowly to her right side and touch her right hip joint. "What do you experience?"

"Yellow and black." I feel a shift in her right hip joint, a loosening. "The black is trying to take over the yellow," she continues. "There's a struggle between the colors."

Often clients come up with colors spontaneously, so I was not surprised when Sweetie did.

I slip the palm of my hand under her shoulder and wait. I feel a tingling; we're on the right track. I ask her about this struggle between the colors.

"Yellow is where I want to be and black is where I've been. Been black inside for a long time."

"Imagine your journey of black changing to sunshine yellow." Her shoulder tightens. I can feel her fear. I know she needs a totally understanding ally who won't hurt her. "Would you like to have allies with you on this journey?"

"No! I have to do it by myself."

"Do you feel better by yourself? Wouldn't you like to have someone with you?"

"I want allies who understand because I hurt too much when they don't."

And so together we create an imaginary ally, full of understanding, empathy, wisdom, grace and peace.

"What kind of outfit does this ally wear?" I ask. The question is meant to be light, humorous. But she takes it seriously.

"Yellow with purple."

Already I see that the black has changed to purple. It is an encouraging sign. "He or she?"

"Both a he and a she—strong qualities of man and gentleness of woman." I can feel her body tremble. She is excited by her new ally. Her eyes are closed. I move my hands down over her arms and legs, feeling some energy shift with my motion.

"Ah," I repeat. "Both a strong man and a gentle woman."

Her trembling increases. "My heart hurts," she says, tearing.

"Put your hand where it hurts."

She gingerly puts her hand on her chest. "You can come out now," she says in a young, whispered voice. "So much violence. So much incest."

This is dark territory. I know that she has been dealing with these feelings for years with several therapists. Through my gentle touch, she has entered an old family trance of aloneness and abuse.

"I have no mommy. No protection." She crunches up into a ball. "Don't want to hide anymore. Not dirty, not shameful. Want to be clean," she sobs, gulping air. My hands softly touch her back. Energy is pouring through her in a definite rhythm.

"Together, how shall we clear this dirty black color that has nothing to do with you?" I ask.

There is a moment of silence, then she declares that the dirty black color belongs to other people, not to her. From a tight ball, she unfolds and lies on her back, telling me she is now willing to send the black color out her fingertips. She points toward the windows. "Help me," she moans.

I pass my hands swiftly up from her pelvis to her chest and out her arms and fingers, not touching her but working with her energy field. The group members moan with her. They are fully focused on Sweetie and me.

I encourage her to move her black energy. "Keep sending it right out. Through the window."

"I'm a survivor!" she shouts. I know that for Sweetie this is a key phrase, something I try to elicit from my clients.

"Yes you are," I say, slipping my hands under her shoulders again. "I sense more than a survivor."

"I'm alive, not just surviving. I'm *clean!*"

"You sure directed a major cleanup job! You deserve peace and yellow, to dance and be part of the circle of life."

"Instead of being outside looking in, I can be inside," she announces, reclaiming her body.

I ask her to feel the yellow again.

"It's the color of love," she says.

"Have it talk to you. Have it say, 'Hello, Sweetie, I love you.' "

"Today is my birthday," she announces.

"Then have the yellow say 'Happy birthday' too."

" 'Hello, Sweetie. Happy birthday. I love you,' " she says. Her eyes are still closed. She is oblivious to the others in the room.

"Say it again," I urge her, keeping my hands under her shoulders. The trembling has stopped. She is still tense, but more peaceful. She opens her soft eyes and gazes into mine. I ask her to sit up slowly.

Blinking, she looks around the room, slides off the table and stands.

"Make contact with anyone and share your key phrases," I tell her.

"I'm more than a survivor. I'm love and today's my birthday." She moves to several people and shakes hands with them. "It's my birthday."

Some people have been crying. They hug and kiss her. We all sing, "Happy birthday, Sweetie."

I play a special song for her on the tape recorder. Music fills the room; the group begins to dance. Sweetie is in the middle of their circle, and for a moment I fear she's going to feel trapped. But no, she begins to dance too, a sunbeam. Laughing and crying, she moves as a free spirit. For now, she is connected.

Upon reflection, there was no question in my mind and body that Sweetie would eventually be able to delve into her deep wounds of abuse and isolation, because she experienced my gentle touch/contact, felt safe and protected and was totally accepted. She had been to a special place, one with significant colors that became a metaphor for her yearning to shift from darkness to light. Now that she had been there, the next steps would be easier.

My "Discovery" of Trance

When I was starting my work in the 1960s, each time I introduced my "listening touch," I noticed a consistent shift in people's bodies and in their tone of voice, eyes and facial expressions. They seemed to go into a special "bubble"—forgetting the group, still hearing my voice, feeling my touch, noticing sensations and moving more deeply inward on their psychic journey. Sweetie's face, for example, became more serene; her eyes closed, her eyelids fluttered, her muscles relaxed, her voice softened and flattened—all signs of a deep inner experience.

It was Jane Parsons-Fein, an Ericksonian hypnotherapist, who gave me a name for what they were experiencing. Jane invited me to present a Rubenfeld Synergy workshop at the New York Society for Ericksonian Hypnotherapy. She asked me if I had ever studied with Milton Erickson. I shook my head. At that time, I hadn't even heard of him, and I was not aware of the way he used language for what I was doing.

After observing several Rubenfeld Synergy sessions, Jane concluded that listening touch, combined with talk, seemed to reassure clients that they were safe and created a direct dialogue with the unconscious mind. This seamless integration encouraged intuition and invited clients into what Milton Erickson called a therapeutic trance.

Curious about trance and altered states of consciousness, I proceeded to investigate.

What Is Trance?

Trance is a state between being fully awake and fast asleep. There are physiological manifestations of it in the brain—as you move more deeply into a trance state, the rate of brain waves decreases—and it can be light, medium or deep. It is not some exotic state, but a normal, everyday experience.

Have you ever sat at a boring meeting and let your mind wander off to another place—a problem, a loved one, the ocean, sunshine? (This is called dissociation.) Remember listening to someone talk on and on until your eyes glazed? Do you recall the feeling you have when you are deep in prayer or meditation? Have you ever been so involved with a task that time seemed to shrink and what had in fact taken hours felt like minutes? (This is called time distortion.) Athletes are especially aware of this phenomenon. In describing his winning basket in a championship game, Michael Jordan reported that he seemed to have all the time in the world to make it, that he could see the entire court and everything that was happening on it, and *knew* he would make the crucial play.

These are all light trance states; everyone goes in and out of them. They are everyday, normal events during which you are still in control—aware of your environment, hearing sounds, smelling aromas, seeing colors—yet you are also in an altered state of consciousness that is very private.

Trance states, whether light or deep, in and of themselves are neither good nor bad. They are transient, occur all the time and are a natural part of living. They can be used as fodder for creativity and change. However, when your trance state is habitual and unconscious, you lose the choice of going in and out of this altered state of consciousness, and your behavior may be adversely affected.

Altered States of Consciousness

Milton Erickson demonstrated the use of everyday language and trance as an entrance to an altered state of consciousness. Experiences and messages stored in the unconscious mind are imprinted in the body. When you close your eyes and imagine a scene, your body will respond as if you are actually there. In this "imaging trance" state, you have a unique opportunity to rescript an event that was stressful, painful or uncomfortable—which is why trance is so important a therapeutic tool. When you remember the warm, loving touch of your partner, your body will also remember. You may even "feel" as if the experience were happening right now.

A baby's nervous system and brain are like sponges, taking in everything without discrimination. For the infant, the neocortex is incidental to sensations, feelings and emotions which are all present and available. Behaviors are not understood in a rational way; they are experienced. Facial expressions convey the meanings behind the not-yet-understood words they are hearing. In these early years, the unconscious, nonlogical brain predominates. In fact, babies and young children go in and out of trance states so quickly they don't know the difference. Children's imaginations are vivid and alive; they "believe" the stories they hear, believe that your nonverbal reactions to them, and what you say about them, are accurate descriptions of how and who they are.

When we grow up, as adults we use our neocortex to distinguish fantasy from reality and to impose order on the world around us. But as we've seen, we are still open to trance states.

I invite you to practice this light trance exercise.

EXERCISE: AN INVITATION TO A TRANCE (SOLO)

1. Find a comfortable place to sit or lie down (you deserve to take some time to experiment, learn and pay attention to yourself). Take a few breaths, allowing each breath to become deeper and fuller. Slowly let your eyelids close, without any effort. Continue to take full breaths as your ribs expand to welcome the new nourishing air and expel the old air. Breathe in nourishment and breathe out stale air, like life's rhythm, ever changing from moment to moment. Rub your hands together. Float your arms apart, about shoulder width, with your palms facing each other. (I got this image from watching my father playing an accordion.)

2. Imagine you are holding an accordion. Your hands fit comfortably on both sides. You can pull your hands apart, creating a large space and a long sound, or squeeze your accordion together, creating a small space and a short sound. Allow your hands to move slowly through the air, closer and closer to each other. Now breathe in, letting new air nourish you. Breathe out, letting go of the old air that needs to leave and making space for new experiences and discoveries.

3. Focusing on your hands moving toward each other, you may hear a distant sound, or may smell and even taste food being prepared. Perhaps you feel warmth or coolness. Integrate these or any other sensory experiences into your journey as you move closer and closer to making hand contact. Notice any changes in density between your hands as you move them.

4. Once your hands touch, take your time and explore them: skin texture, temperature, sensations and movement.

5. Your eyelids are still closed. With the fingertips of your right hand, gently explore the crevices between the fingers of your left hand. You may be surprised how sensitive your fingers are as they move over the outside, the palm and the wrist of your hand.

6. Let your right hand rest. Move your left across, underneath and between your right fingers.

7. Slowly separate your hands and place them on your lap, allowing yourself to rest. Now enjoy the contemplation of a pleasing image–maybe the oasis of a calm lake surrounded by a leafy forest. Perhaps you'd like to rest on a beach, listening to the waves roll in and out in time with your breath. The sun shines as you lie on the sand, warming every bone in your body, radiating heat and pleasure. It slowly begins to set, signaling that it's time to go home. You brush the sand off, roll over and slowly stand up. Fresh, salty air touches your tongue, a coolness rushes to your nostrils and the waves play a steady sound. You walk home, allowing your feet to sink into the sand easefully, feeling satisfied and full. Every cell in your body knows that you can return to your oasis anytime you want, to reexperience relaxation, health and pleasure. Arriving back from the journey is easy. Your key fits perfectly into the lock. The door and your eyes slowly open. Your friends and family are there, smiling and welcoming you back from your trip.

8. Glance around the room. What do you see? How bright are the colors? What sounds, smells and textures do you experience?

The Family Trance

"The deepest trance you'll ever be in is your family trance," says Jane Parsons-Fein. When you think you've changed (whether thanks to therapy, support groups, workshops or simply increased self-knowledge), just go home for the Thanksgiving or Christmas holidays and experience those old emotional buttons being pressed by family members. You are back in a trance which started the moment you were born.

When baby Adam entered the world he immediately became part of a family trance. His energy field was open, highly permeable. Unconsciously and consciously he absorbed and experienced everything around him—vocal tones, aromas, facial expressions and ways of touching. These family interac-

tions became a vital aspect of his character and personality. Impo
tive elements came soon after—the introjections which make us f
smart, winners or losers, etc.

The *memory* of the people in any trance state—parents, grandparents, siblings, aunts and uncles—can trigger emotions connected to them. Before you can rationally understand why you feel so emotional, your body is already swimming with their words and behaviors.

In my early Alexander lessons with Judy Leibowitz, the touch I experienced evoked early memories of poverty, my absent father, my harassed mother, the death of family members. These were already embedded in my body. I'm sure my teacher did not set out to create this altered state of consciousness, but unwittingly she opened my body and mind to these extraordinary experiences.

There are, of course, loving and warm family trances as well as unsettling ones. I recall the delicious feeling of having my back gently stroked by my mother. It was calming then, inviting me into a relaxed trance. It is still calming today.

As a little experiment, when you're agitated or stressed, slowly rub a piece of velvet with your fingertips. You may find that your breath slows and deepens as you begin to relax, and that you become calmer and quiet; your body thanks you for slowing down and resting. Perhaps touching the velvet reminds you of your favorite fuzzy animal and your "blanky," that sacred, tattered piece of cloth you needed so badly in childhood.

The Dysfunctional or Symptomatic Trance

Early traumas are fertile fields for habitual dysfunctional and symptomatic trances, which are similar to addictions. People may know that their behavior is alienating others or themselves, yet they do not stop. These symptomatic trances begin unconsciously early in life and are rationalized as "normal" by the neocortex. Sweetie lived in a "fear trance" from her early childhood into her present adult life. No matter how much nourishment and joy surrounded her, she was too afraid to reach out for it. Early abusive contact created fear in every cell of her body. In this traumatized state, she believed people would harm her, and no rational thought could change her deep, symptomatic trance of terror. She watched life from the sidelines. It was only at the workshop that she summoned up the courage to travel into a therapeutic trance—the first, I hope, of many.

The Therapeutic Trance

In this trance, you journey into the unconscious landscapes of your mind, where you can unpack your baggage of attachments to a particular way of being.

One such attachment is a distorted body image. "You're too fat." "You're too thin." "Your breasts are too big/too small." "You're too short/too tall." Imagine the trauma adolescents go through as their bodies change. Verbal and nonverbal reactions to these changes can contribute to a dysfunctional body trance that may persist into adulthood and create serious self-esteem problems.

Touch is especially effective for distorted body trances. A thin and svelte woman perceived herself as fat. She suffered through bouts of bulimia and anorexia. During a session I touched her shoulder and remarked that she had strong, substantial bones. I invited her to touch herself. Her eyes widened in surprise. "I'm not fat or big at all!" she exclaimed. "I just have large bones."

In therapeutic trance, clients often move deeply inward, where they may have "inner conversations" that are not expressed vocally. It is at these times that touch maintains contact and sustains a lifeline between the client and his/her Synergist. In this altered state of consciousness all sensory channels become highly sensitive, especially the sense of sound. The speaker's voice (sound and rhythm) is partnered with touch in a harmonious and meaningful way. Vocal tones themselves have as much meaning as the uttered words behind them.

EXERCISE: SOUNDS WITHOUT WORDS (SOLO)

Here's an exercise in which gibberish has meaning.

1. Stand in front of a mirror and tell your image an emotional story—without words. Make any sounds, gestures and movements you need to tell the story, as if you were speaking an unknown foreign language.

2. Complete your story and take a few deep breaths. Close your eyes and travel inward. What are you aware of now? How did your body experience your story? Were there parts of the story that were easier to express than others? How much movement did you use?

EXERCISE: GIBBERISH (DUET)

1. Introduce yourself and, without using words, tell your partner about some aspect of your life. Explore the whole range of your voice, from a low pitch to a high tone. Let your arms, hands and body move. Notice your partner's facial expression as well as your own body experience.

2. When you complete your story, you both may burst out laughing, or you may look serious. Ask your partner to describe, in words, what he or she imagined you said.

3. Check out how much emotional content came through to your partner, even though you did not use words.

4. Both of you close your eyes, take a few breaths and travel inward. Allow yourselves to experience any thoughts, feelings and sensations you have, without doing anything about them. A few moments will be all the time you need.

5. Slowly open your eyes and softly glance at one another. Did you experience knowing each other in a different way? Share anything you wish about this "knowing." Were the speaker's vocal tones and facial expressions congruent with the import of the story? You heard the story on a "gut" level, not with your rational mind, and perhaps it made a great deal of sense.

6. Switch roles. Follow Steps 1–5.

Vocal Double Messages

You can also experiment with a vocal tone that does not match the content—for instance, "I'm so happy" said in a slow, flat monotone, or "I'm really worried" said in a fast, high-pitched, eyebrows-up tone, or "You seem sad" said in a bright, chipper voice.

Take a moment and try these vocal double messages. Notice how your body reacts to each one. Where does your body tell the truth as you say them?

In Rubenfeld Synergy sessions, it is the Synergist's intention to make sure that what is being said, in both words and tone, matches the emotions that are felt in the body; otherwise clients may be jarred out of their therapeutic trance. People are keenly aware that the Synergist is listening to them on all levels. Because of this special congruent partnership, clients experience a "healing trance" that is rich with metaphors.

Metaphors in Therapeutic Trance

Clients' life stories are expressed through their bodies' metaphors, which become powerful means of self-discovery. For example, after touching Sweetie's left hip, I asked her to talk to it. In a fully awakened (nontrance) state, the neocortex would have said, "Talk to the left hip? Are you crazy? A left hip doesn't carry on conversations." But in a therapeutic trance, the rational mind takes a vacation while the intuitive, emotional, nonlinear sensory brain, accompanied by trust and a feeling of safety, understands these suggestions as naturally making sense. When Sweetie said, "Relax," I felt her body tighten, for she was still afraid of her feelings and could not follow her own verbal command. But when she shifted to the metaphor "color purple" her body softened, and I knew I could help her "relax" through the use of color. One of her sides was purple, and the other became yellow and black. Realistically, pelvises and hip joints don't turn into a variety of colors, but in an altered state, everything is possible, and I proceeded to develop her life of yellow and black as metaphors for Sweetie's wounded psyche.

Often, workshop participants say they have "seen their session," even though they were not on the table. They have recognized their story through the universal metaphors of others' experiences.

I asked a group to try a short experiment. I invited everyone to close their eyes and imagine a scene in nature. When it was time to return and open their eyes, one of the participants, Catherine, was crying. She had seen an apple fall to the ground, not far from its tree. She sobbed as she recounted how the apple lay there. I walked over to her and touched her back. It was tight and unmoving. I asked her to "become" the apple and retell the story in the present tense.

She closed her eyes, beginning to enter a trance state. "I am an apple falling to the ground, not too far from my family." My hands were still on her back, and I could feel her shiver.

I waited a moment for her trance to deepen. "And then what happens?"

"I go into the soil and decompose—fall apart." Her shivering increased.

"And then?"

"It's winter and snow covers me." I could feel her body become flaccid.

"And now?"

"Spring awakens and revitalizes me. I come through the soil as another apple—full of nourishment, juicy, fat, red and *no worms!*" Catherine's body was vibrating with energy and laughter. The group laughed with her.

As Catherine became an apple, her story of fear, falling apart, starting again and reemerging as a "juicy" woman unfolded. In light of it, I want to take you on a short guided fantasy.

EXERCISE: THE ROSEBUSH METAPHOR (SOLO)

1. Find a comfortable place to sit or lie down. Have some paper, crayons and Magic Markers nearby. When you are settled, close your eyes and take a few deep breaths, letting go of any tensions.

2. Imagine a rosebush. Now allow yourself to "become" the rosebush. Notice how you feel. What is it like to "be" the rosebush?

3. As the rosebush, look around at your environment. Where are you growing? What kind of ground do you come from? Feel your stems and branches.

4. Where are your roots? Are you a single rosebush or are you part of a community of rosebushes?

5. Who planted you? Or are you growing wild? How do you experience winter, spring, summer and fall?

6. Does anyone touch you, smell you or look at you admiringly? Does someone feed you? How have you changed over the years?

Take all the time you want to explore the images, feelings, sensations and emotions that come up. When you feel complete, take a deep breath and slowly open your eyes. Right now may be a good time to write or to draw yourself as a rosebush.

You may not feel much connection with the rosebush metaphor. If not, pick another. Each individual is unique. It is from one another's stories that we create a therapeutic trance, for each of our stories is also a universal metaphor. The suggestions of images and metaphors are built around these stories, not imposed by the Synergist, whose job is to act on what clients recount.

During a session, everyone may become participants in the therapeutic trance state and use the metaphors for themselves. The connections established between people in this state are powerful, regenerative and heart-opening.

The Paradoxical Trance

What an extraordinary training I received to become a music conductor. I had to learn and practice how to focus my attention on details (individual notes, phrases, dynamics) while simultaneously holding the bigger picture of the entire piece in my head.

This is called a paradoxical trance. Music conductors and Synergists share the ability to be comfortable with fluctuating between inner experiences and outer awareness. While conducting an opera, I had to hear all the parts internally, listen to the musicians and remember that the tonal "painting" was larger than the sum of its parts.

The Meditative Trance

Some exercises that induce altered states of consciousness are called meditations. Imagery, prayer and chanting are used to enter a relaxed, peaceful state of being and oneness with the universe.

When you close your eyes, take a few deep breaths and travel inward, you are moving into a meditative trance state. Many exercises and practices begin this way so that you can slow down, become more aware of your thoughts and movements and eventually release them.

I strongly advise you to meditate as often as possible—every day if you can. You will be amazed at the transformation it brings.

From Trance-Formation to Trance-Reformation

Let's go back to Sweetie's story. With a minimum of personal history, we moved to a very early "black" trance-formation. ("Been black inside for a long time.") In her use of the black metaphor, Sweetie was telling us that her wounds were deep and her isolation severe. Instead of healthy contact, she experienced abuse trauma, which created terror. ("Yellow is where I want to be and black is where I've been.")

Because she was in a therapeutic trance, feeling safe and experiencing positive touch/contact, she was able to express her yearning to move from darkness into light. At this moment, Sweetie was in the process of *trance-reformation.* By substituting a positive experience for a negative one, she was able to integrate a changed self-image, to go from a "dirty little girl" trance to a "clean little girl" trance. From a woman too traumatized to join a group to a woman capable of dancing in the center of that group.

Her euphoria was a joy to experience that day, but she will probably have to walk through the tunnel again and again—accompanied, one hopes, by loving and supporting allies. Her wound was too deep to be healed in a single session, no matter how profound.

By guiding clients into therapeutic trance states, Synergists help them deal with their problems on an unconscious level. We return to those early *trance-formations* where their problems began. In order to change, people need to detach from their earlier trance states and imagine themselves surviving in life without their dysfunctional trances. Fear and anxiety may be the driving forces in their behavior. By facing and rescripting body/mind memories, they emerge into a *trance-reformation* with new awareness and different behaviors.

Clients show us caregivers their face of trust. They share their life themes and variations; they are emotionally naked. We listen with our eyes, ears, hands and hearts, thus enabling us to guide them on a path of learning and growth. And throughout we listen to ourselves. For the first step in caring for others, something that professional therapists and laymen alike too often forget, is self-care.

10.

The Art of Self-Care: Who Heals the Helper?

Richard, a psychotherapist, complained to me in a private session that after seeing his clients, he would experience stiffness and sharp neck pain and feel drained and tired. When I tried to move his head and neck I found them frozen—stuck together. "Show me how you sit and listen to your clients," I asked. He assumed his habitual "listening" attitude—neck stuck out, back rounded, chin jutting forward. The position, he believed, conveyed contact, concern and total focus. He was literally "sticking his neck out" for others.

In reality, his posture did the opposite. It gave him terrible pain, making it virtually impossible for him to concentrate and remain present with his clients. This habitual posture was not new, he told me; I had a hunch it had started when he was a child. When he was able to relax his neck, he became anxious and sad. Through questioning, I led him back to his childhood, and he remembered going into this trance-fixed posture whenever he heard his younger brother and sister cry. He would rush to take care of them and forget all about himself. For him, loving and empathizing meant giving up his own physical comfort. This deep, early trance state had not changed, and now in his adult life he still listened to others in the same way.

Richard, a caregiver, needed to learn to care for himself. Over the next few months, he became less attached to his "superempathizer" role and was willing to experiment with more comfortable postures.

Self-Care Is for Everyone

When I began giving workshops on self-care over thirty years ago, I intended them primarily for professional therapists such as Richard. I quickly realized, however, how important self-care is for anyone working with and caring for others: healers, doctors, nurses, counselors, social workers, performing artists and especially parents and other family caregivers. A powerful theme began to surface: how to continue doing your work, raising a family or caring for the elderly without burning out.

Many workers complain of stress at their jobs, and stress-related illnesses are increasing at an alarming rate. There is always some degree of stress in life. The trick is how you respond to it; how it affects you emotionally and physically; how quickly you recognize its signals. When a cluster of stressful situations evolve into *distress,* your health is greatly compromised.

The Listening Touch Demands Self-Care

Self-care is continually woven into the fabric of Rubenfeld Synergy. It is part of my basic philosophy and is highlighted throughout the training of Synergists.

Through using listening touch, Synergists become vulnerable to the physical aches and pains of their clients. The reverse is also true: practitioners may communicate their tensions, problems and physical imbalances to their clients.

Rubenfeld Synergy focuses on the self-care of both client and practitioner. When professionals continually feel drained after sessions, they will eventually suffer from exhaustion. Thus physical and emotional self-care is crucial in preventing burnout.

Symptoms of Burnout

Burnout has become a catchall phrase for the dangers awaiting those who give too much. In the art of self-care, people learn to listen to themselves and to the warning signals that demand attention before exhaustion and illness set in.

The burnout phenomenon is often accompanied by fatigue, body aches, insomnia and a sense of depletion. Most of all, burnout involves a feeling of being drained rather than energized by work and life. I've supervised many health professionals, and I've noticed time and again the same warning signs. See if they are familiar to you as well:

- Are you a member of the Helpers' Club? Does your overzealous helpfulness blur your boundaries? Do you absorb the physical and emotional problems of your colleagues, clients, friends and family members? Do you have difficulties resisting the "urge to merge"?
- Do you feel *drained* after meeting with certain types of people? Where do you notice tired feelings in your body? Are you aware of those toxic people who absorb energy like sponges?
- Where are you on the care priority list—first, middle or last? Are you like the cobbler who makes shoes for everyone but himself?
- Do you have a *superwoman/superman* complex, taking on too many tasks and not delegating to others? Are you a "supercontroller" who is convinced that unless *you* do the job it won't be done right?
- Are you a workaholic? According to Jeanne Hanson and Patricia Marks in their delightful book, *You Know You're a Workaholic* when: "You prefer Mondays to Fridays"; "You hope that if you come back to life as an animal, it will be an octopus"; "You rent movies you've already seen so you can work while watching"; "You write love letters on Post-its"; "You have an anxiety attack every time you finish a project"; "The first eight numbers on your speed dialer are restaurants that deliver"; "You keep more notepaper than toilet paper in your bathroom"; "Your favorite foods include Jiffy Pop, Minute Rice and Instant Breakfast"; "You present the minister with an annotated plan for paring down the moment of silence."
- Do you make lists of all the "shoulds" you have to do? If so, have you made a list of "wants" lately?
- Do you fear you'll be a "bag lady/bag man" unless you're working on many projects at the same time? Will you hear a voice telling you you're "lazy" unless you keep on working?
- Does every detail become so time-consuming and demanding that you can't carry through on your plan?
- Do you think that unless you've accomplished your task perfectly, it isn't

worth much? (A particular failing of musicians. There's always something that wasn't played just right.)

These are only a sampling of a longer list. If you've said "yes" to any of them, you're ready to begin your journey of self-care. Indeed, how you care for yourself may save your life *and* the lives of people you love and give care to. Although it may appear paradoxical, the more you care for yourself, the more health, energy and emotional vitality you will have left to care for others. By continually filling up, you can spill over without depleting yourself.

Self-care means exactly what it says: you are in charge of your own care. The following exercise should be done in private. (You don't need a Synergist to help you!)

EXERCISE:
A SELF-CARE GUIDED JOURNEY (SOLO)

Find a quiet place, sit comfortably and close your eyes. Travel inside your body. Return to a time when you became aware of taking care of yourself physically, emotionally and spiritually. Perhaps you were little and found out what you needed to do in a difficult situation. Maybe, older, you discovered how to take care of yourself in the face of a challenge. Perhaps there were adults in your life who modeled the ability to healthfully care for themselves. They were capable of listening to you and knew when it was time for them to stop and listen to themselves. They were clear about their boundaries.

Allow feelings of pleasure and satisfaction to permeate every cell in your body, knowing you have self-care allies. Bring them along with you as you slowly open your eyes and return to the present. Take some sheets of paper and write down anything you wish about this experience.

Questions to Guide You:

- How did you experience your body during this journey?

- What scene did you remember?
- Who were the main characters?
- How did you take care of yourself?

Maggie: Finding Life Force

Maggie, a nun, struggled for years with the question of whether to stay with or leave her religious community. I asked her how her body experienced her indecision. "I was so tired," she told me. "I couldn't work and I hardly moved. Every part of me hurt."

Maggie's body was screaming the solution. I didn't touch her, but I could see it. "I finally left, and soon after, my energy returned. I realized what I needed and sought professional counseling." When you listen to your body, it becomes a metaphor for what is happening in your life right now.

Ruth: Letting Go of Superwoman

Ruth, a young home owner, had problems with her kitchen drain. It drove her crazy. Usually, she'd be obsessed with fixing it herself and would stay up half the night trying one thing after another. But this time, after she failed on the first try, she decided to take a long hot bath. The next day she got a plumber (lucky lady!) who was able to solve the problem.

From Ruth's simple story, we can extrapolate seven steps that she used for self-care.

1. Recognize a stressful situation.
2. Let go of habitual behavior (in Ruth's case, trying to fix the problem herself).
3. Listen to the body's pleas for relaxation.
4. Detach from the need for immediate gratification and wait till another time.
5. Have faith that the solution will happen.
6. Have a backup plan (in case that plumber wasn't available, Ruth had compiled a list of others).
7. Give up the Superwoman/Superman trance.

Ruth handled her anger and frustration in a healthy way and diminished distress in her life. Use these seven steps the next time you face a similar crisis.

Donna: "You're Not My Mother"

During her "Self-Care Guided Journey" at a workshop, Donna touched her neck and throat, her face red, then shared in a constricted voice her story of graduate school troubles. She had applied for an internship and been rejected because the department head, Sally, wanted a man. "I could hardly swallow or eat," she told the group. "I felt like someone was choking me, wringing my neck."

Eventually the decision was reversed. But Donna still didn't trust Sally, whom *she* wanted to strangle. Instead, she had begun to choke herself. I quickly intervened and touched her neck. It was rigid.

I gave her a towel to use in place of Sally, and asked her to twist and wring it and imagine she was choking the department head. This was a novel idea for her. She tried it, and I could see her anger begin to dissipate.

Although Donna's neck was softer when I touched it again, her back was still tight. I asked her if she felt complete. She shook her head, muttered something about being lied to and began to cry.

"It's painful being lied to—let your sadness flow through," I said, placing my right hand lightly over her stomach and my left hand on her upper chest. Sadness poured out from her chest. As her story unfolded, she told me that her mother was highly respected, considered perfect, loved by her community.

"How did she become 'not perfect' for you?" I asked, moving my hand on her back and asking for the opposite of her original description of her mother (a therapeutic intervention I use frequently).

Her expression became angry. "At home she could be mean and cruel, so I didn't trust her."

"Say that phrase to your imaginary mother. 'You can be mean and cruel.' "

"You can be mean and cruel," she said several times.

"Say it again," I told her, "but this time try adding the phrase, 'And you're still my mother.' "

Both statements were true: her mother could be mean and cruel, and she was still Donna's mother, someone Donna loved. My purpose was to get her to accept the paradox.

"You can be mean and cruel, and you're still my mother," Donna repeated. Her back softened dramatically as she accepted and integrated the two statements. Now I made a leap and took her back to her work situation. I asked her to say these words to the department head, changing them in any way she wanted. "You can be mean, cruel, lie to me—and you're still my supervisor," she said.

I asked her to add, "And you're not my mother!"

She repeated that and paused for a few moments, looking surprised. "I feel free and unblocked now," she said. "I didn't know I was carrying so much anger."

Once again, we can derive seven universal self-care principles from her story.

1. Listen to signals in your body.
2. Give your body a voice and let it speak to you.
3. When you hurt yourself, ask if you really want to hurt someone else.
4. Redirect your anger and frustrations. Hit a pillow or twist a towel instead of hurting yourself.
5. Are you projecting some unfinished business with family members onto other people?
6. Imagine your supervisor (or anyone else) and separate him/her from your family. Say, "You're not my (name of family member)."
7. Listen to your body and notice any physical and emotional changes now.

Waking Up to My Own Self-Care

Many people begin their journey of self-awareness and self-care during a physical or emotional crisis. As a teenager I practiced the viola for hours a day, not realizing that my hunching over, twisting and pushing myself through discomfort and fatigue would lead to serious physical problems later in life. Conducting exacerbated my condition. It took a screaming back spasm to wake me up and prompt me to begin my self-care journey. I didn't even know what relaxation was, let alone how to play music easefully.

I searched for books about relaxation and found a 1930s title, *You Must Relax*. Laughing at the stern, commanding tone of the title—who could relax under such pressure?—I read it through, discovering that the state of relaxation is difficult to describe in words and even more challenging to experience.

Send It to Your Angels

Another healing crisis overwhelmed me in the 1970s, when my schedule was jam-packed with people who depended on me to ease their emotional and physical problems. With an empathetic open heart, I gave my all to help them pass through their painful and difficult dilemmas.

My right fingers began hurting. But my intent to help others was so strong that I overrode signs of my own pain. "The show must go on no matter what" had been my credo as a musical performer. Now here I was, following the same "performer trance" with my clients and workshops. When the pain became too strong to ignore, I stopped long enough to assess the situation, and decided to suspend the use of touch temporarily.

Pierre, a friend from Brazil, invited me to give some workshops in his country, and agreed to my proviso that I not use touch. A few weeks into my stay, he took me to a small village that was famous for the way its traditional healing practices had been combined with Catholicism. We entered a church full of worshipers, dancing the samba while chanting "Kyrie Eleison." Not your ordinary church service, I thought, and with joy imagined parishioners dancing around the nave of St. Patrick's Cathedral in New York.

When the mass ended, people lined up to be healed. White-robed male and female healers appeared in every doorway. They puffed cigars and blew smoke all over the sick ones. I don't smoke, and as my throat constricted from the fumes, I could feel my panic rising. Still, I watched closely as the healers held their hands an inch or so from the sick people's bodies and made a kind of wiping or brushing movement, as though they were removing dust or cobwebs from their clothes. Then they shook their hands out as if they were flicking drops of water into the air. Between the humid air and the smoke, I began to feel faint. A tall priest appeared at the head of the nave, wearing the Star of David, the Moslem half-moon, a Catholic cross, Celtic designs and other symbols on a gold chain around his neck. At the appearance of this ecumenical being, the healing ritual suddenly stopped. The church became bone quiet.

"Will the Tree come forward," he proclaimed in majestic Portuguese. I looked around. No one moved. Pierre translated.

In Hebrew, "Ilana" means "young tree of life," and somehow the priest knew my name and, I gathered, was referring to me. But there was no way I was going down that aisle. Pierre took my arm and pulled me forward, and there I stood in front of this six-foot-two figure. He peered down. "You are a healer," he said softly, "but you are only one little woman and you are trying to heal everyone."

Tears flooded my face. He took one of his religious emblems and touched me at several places along my spine. "Here in Brazil, everyone sends their illnesses, pains and problems to their angels," he said. "Yours are unemployed, just waiting for you to send all those problems to them."

I was in a trance state. My rational "neocortex mind" was flashing signs of disbelief, while my limbic brain intuitively knew exactly what he meant. Confused, I returned to my seat. Hours may have passed. The singing and dancing resumed, but the music seemed far away. I was revisiting my childhood and remembering all those times I took care of my parents, at the same time feeling deeply moved and grateful to the priest and the ritual.

Slowly I returned to awareness of the church, Pierre and the dancing people. I felt a mixture of pleasure, awe, amusement, mystification and release. I had experienced a cellular trance-reformation, as if those early messages had reshuffled themselves and were now in a more balanced energy, creating space for me to heal myself.

When I got back to my practice, I experimented with the brushing-off movement the Brazilians had used. I still didn't understand its effects scientifically, but it seemed to work, and I've used it to this day.

Now, if I'm asked why I shake my hand out in a brushing-off motion, I tell the Brazilian Angel story. You can learn the "brush-off" through the following exercise:

EXERCISE: TOXIC BRUSH-OFF (SOLO)

Close your eyes and imagine encounters with toxic people. They've been complaining, not listening, absorbed with themselves, judgmental. Nothing you can say or do is right. When they leave the room, you may feel drained. Take a deep breath and follow these steps:

1. Rub your hands together and shake them out. Begin moving them in

short, feathery brush strokes (letting your fingers lightly touch your clothing) up from your pelvis, stomach, chest, neck, face, head and into the air above. Shake your hands out.

2. Focus on your hands. With your left hand, brush down from your right shoulder, your upper arm, elbow, forearm, wrist, right hand, fingertips and out into the air. Shake your left hand out.

3. Now, with your right hand, brush down from your left shoulder, your upper arm, elbow, forearm, wrist, left hand, your fingertips and out into the air. Shake your right hand out. Rest a moment and take a deep breath.

4. With both hands, brush your left leg, starting down from the left hip joint, down your left thigh, knee, calf, ankle, foot, toes and out into the air. Shake both hands out. Pause and take a breath.

5. With both hands, brush your right leg, starting down from the right hip joint, down your right thigh, knee, calf, ankle, foot, toes and out into the air. Shake both your hands out. Pause and take a deep breath.

When brushing off toxic people, you might want to name them and say: "I'm letting you go" or "I release you" or "I don't want your toxicity"—whatever phrase you wish.

When you've completed the exercise, listen to your body. What are you aware of now? How do you feel now as compared with during the moments after they'd just left?

The exercise works well after receiving a toxic phone call, fax or E-mail. If you have time, write about any images, feelings, thoughts and physical sensations in your journal.

EXERCISE: TOXIC BRUSH-OFF (DUET)

1. Stand in front of your partner, who will be your "brusher." Close your eyes and imagine a toxic person in your life. Open your eyes and share anything you wish with your partner.

2. Close your eyes again and take a deep breath.

3. The brusher, facing you, rubs his/her hands, shakes them out and begins light, feathery brush-off movements up from your pelvis, stomach, chest, neck, face, head and out into the air.

4. Now, your brusher-partner moves around, facing your back, and begins brush-off movements up from your buttocks, lower, middle and upper back, neck, back of head, top of head and into the air.

5. The brusher moves to the side, facing your left shoulder. He/she gently places one palm in front of your chest, the other in back of your left shoulder blade. With both hands, your partner brushes from your left shoulder, down your upper arm, elbow, forearm, wrist, fingertips and out into the air. Now is the time your partner needs to shake his/her hands out vigorously in a direction away from you.

6. Your brusher moves to the side, facing your right shoulder. He/she gently places one palm in front of your chest, the other in back of your right shoulder blade. With both hands, your partner brushes from your right shoulder, down your upper arm, elbow, forearm, wrist, hand, fingertips and out into the air. Again, your partner needs to shake his/her hands out vigorously in a direction away from you. You might name a toxic person and repeat the phrases you used when you did the Toxic Brush-off exercise alone.

7. Your brusher moves around and faces you from the front. Slowly open your eyes and gaze at your partner. Share anything you wish with each other. You may both want to write about your feelings, impressions, sensations, thoughts and images in your journals.

Now switch roles. The person who was "brushed off" will now do the "toxic brushing." Follow Steps 1–7, taking as much time as you want.

GRAND FINALE FOR BOTH SOLO AND DUET

Close your eyes and take several deep breaths. Imagine yourself standing (or sitting) in the center of a golden translucent column. The golden healing energy permeates every cell in your body with clarity and love; it is available at any time and anywhere you wish. Perhaps you see the golden color change to purple or to a turquoise blue. It protects you, surrounds you as you move through unfriendly territory. This is *your* column. Experience how your healing, loving column moves with you. There will be times when you want your healing column immediately. All you have to do is imagine it and it will be there. Your column is full of abundance—always ready to fill you with its love and healing.

Extending Yourself

As we care for others, we extend ourselves. We are literally taught to lean forward, like Richard, as a sign of paying close attention, disregarding the fact that this posture is stressful.

When I supervise therapists, I ask them to be aware of their own movement and bodies. How do you experience yourself physically when you listen to others? What do you notice about your posture, your breathing, your level of physical tension and your energy?

At a recent workshop, Meg, a traditional psychotherapist, was describing her difficulties with a particular client who returned week after week with the same problem. I asked her to re-create the office scene with another participant, who role-played her client. As the volunteer client droned on, Meg, who had been leaning forward to listen, began to sink down into herself. I interrupted and asked Meg to go inside her body and listen to it.

"What do you really want to do?" I asked.

"I've got to move. I want to get up and move around."

I suggested that she do just that.

"I can't. I've been taught to sit in one place and not leave the chair."

"Try it and see what happens."

So she got up and moved around. Then I asked Meg to tell the client how she was experiencing herself and why she'd moved out of the chair. "Even though I want to listen to your story, I'm finding it difficult to concentrate," she said to him. "So I'm going to move around a little."

The volunteer client looked surprised and suddenly perked up. *What's this? Something new is happening!* Now Meg had the client's attention, and the dynamic of the session was changed. Even though the therapist had acted on behalf of her own self-care, she had created an opportunity for surprise and revitalization, for both therapist and client.

Meg and Richard had thought that "sticking their neck out" was the proper listening posture. The following exercise will help you become aware of the relationship between your head and neck and move you through the steps to change it. When the head juts forward, neck and head become a warring couple; when they're aligned, they're loving. When you create this healthy alignment, your spine will lengthen and pitch in to support your head.

EXERCISE: THE RELATIONSHIP OF HEAD AND NECK PART I (SOLO)

Preparation: A quiet space, a long mirror and a straight-backed chair.

1. Stand in front of the chair, feet slightly apart, and sit down. Now, get out of the chair. Repeat sitting down and getting up a few times.

2. Touch the back of your neck with your fingertips. Sit down slowly and pay attention to how your neck feels as you move to sit. Pause for a few moments.

3. Just *think* about getting up, and as you sit let your hands hear how your neck prepares for action. Slowly stand, sensing with your fingertips how your neck moves.

4. Alternate sitting and standing quickly and slowly, exaggerating your movements.

Questions to Guide You:

- Is my neck becoming *shorter* as I move to sit down?
- How do my neck and head feel when I'm sitting still?
- What happens to my neck and head when I *prepare* to get up?
- How do my neck and head move when I actually get up and stand?

Shake your hands out, rest and take a deep breath.

5. Now place your fingertips on the back of your head, with your thumbs at the base of your skull. Sit down and stand up again, letting your hands hear how your head and neck move. Repeat getting in and out of the chair several times until you sense the relationship of your head and neck.

6. Close your eyes as you sit down and get up. Is there any difference in the positions of your head and neck? Now sit down and get up in slow motion, listening to any differences in your head and neck.

Questions to Guide You:

- When do your head and neck crunch together (shorten)?
- When do they stretch (lengthen)?

Shake your hands out and rest.

EXERCISE: THE RELATIONSHIP OF HEAD AND NECK PART II (SOLO)

Can you imagine how many times a day you sit down and get up? See, in your mind's eye, how the vertebrae of your upper spine squeeze together each time you move from standing to sitting and from sitting to standing. The challenge is: how to continue lengthening your neck as you move to sit down and get up.

The relationship of your head and neck are influenced by your eyes. If you look straight ahead while you sit down, your neck will shorten, squeezing the vertebrae together. So let's experiment with your eyes and find out what happens.

7. Touch the back of your neck with your fingertips. Allow your eyes to look downward as you move from standing to sitting. What do you notice about your neck? Did it continue to lengthen?

8. Allow your eyes to slowly move upward as you rise, and look straight ahead once you are standing. What do you notice about your neck this time? Repeat Steps 8 and 9 several times until you are more familiar with this new relationship of head, neck and eyes.

9. Let everything go. Shake your body out and walk around the room. How do you experience your head-and-neck relationship?

Congratulations! You are now ready to apply what you've learned.

EXERCISE: HOW TO STOP STICKING YOUR NECK OUT (SOLO)

Imagine being in the middle of a conversation. You are listening intently and are concerned about what is being said.

1. Lean forward to hear every word. Stop in that position (don't move!) and touch the back of your neck. What are you aware of at this moment? What is the relationship between your head, neck and chest? Shake your hands out.

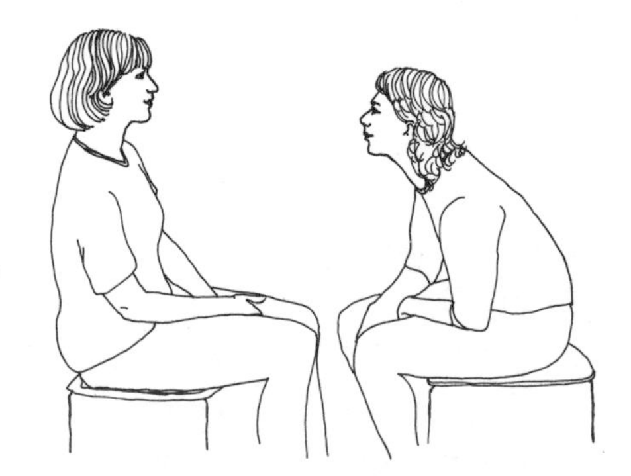

2. Let go of this leaning-forward position and move back onto your chair. Your pelvis is designed to support the weight of your sitting body on its two strong "sitz bones." (Most people are not aware that sitting easefully has to do with their sitz bones.)

3. Slip your left palm under the middle of your left buttock and your right palm under the middle of your right buttock. You will feel your sitz bones clearly. How do you experience them?

4. Lean forward. Stop and notice where your sitz bones are now. Have they moved backward? What do you think you're sitting on instead of your sitz bones? (Answer: your pubic bone.)

5. Lean way back. Stop. Notice how your sitz bones moved and in what direction. What are you sitting on now instead of them? (Answer: your tailbone.) Release this position and return to the middle, sitting equally on your sitz bones.

6. Slide your left palm away from under your left buttock and shake your hand out. Slide your right palm away from under your right buttock and shake your hand out. You are now sitting on your sitz bones again. Can you feel them more clearly? What do you notice about your head, neck and chest when you sit on your sitz bones? They certainly know how to support your spine, create balance and relieve muscle tensions.

7. Now slowly stand, walk around the room and let your body movement sing. You have learned another way of listening to your body that will enhance your ability to take care of yourself.

Self-care involves the physical, emotional and spiritual aspects of your life. All require *awareness,* the essential factor for any change.

Environmental Self-Care

Let's start with light and air.

If I had my way, I'd abolish the use of fluorescent lights, open the windows for "real" air (in sealed high-rises) and eliminate small work cubicles.

What environment do you live and work in? Have you considered that it can be either toxic or nourishing? You may not be able to change the city's air, but you can install air filters. What about lighting? John Ott, creator of full daylight spectrum lightbulbs, showed how schoolchildren calmed and plants grew twice as strong under full spectrum lights.

Spacial environments affect our inner physical and emotional sensibilities. Health professionals suffer from a high incidence of burnout. They face stressful issues daily, and many work in institutions and hospitals that are notorious for small, cramped, airless rooms. Some modern hospitals have created light and colorful rooms for patients, but I remember those old hospital wards, painted a bilious pale green, devoid of paintings or a cheerful, healing atmosphere.

I am especially concerned about the spaces created for Rubenfeld Synergy practice. As opposed to most health clubs and spas (which delegate their smallest cubicles to their bodyworkers), Synergists are encouraged to consider the importance of their professional surroundings, and to create a healthful place for both themselves and their clients. Indeed, your life too will improve immeasurably if you can walk into a space and say, "I'm glad to be living/working here."

As with space, sound can influence well-being. The cacophony of a street or highway can drown out thought, limit concentration, create stress and affect work. Loud music has so damaged people's eardrums that full or partial deafness has become a fact of life. Dr. Alfred Tomatis, a pioneer in audiology, discovered hearing deterioration in aviation employees. He also found that monks became depressed when their order stopped singing their daily Gregorian chants. He established the significance of the ear in synergistic relationship to listening, learning, body image, emotions, voice and language. Billie Thompson, a Rubenfeld Synergist and director of the Listening and Learning Center, introduced me to Dr. Tomatis at one of his seminars. "You use touch the way we use the ear," he told me. "Both listen to the truth of life."

There is no doubt that sound pollution is a factor in burnout. While circumstances may make it impossible for you to escape it, the more protection you can afford yourself (from earplugs to double windows), the better off you will be. As a musician, I am particularly sensitive to sound. I also know the uplifting, soothing and emotional effects of the organized sound of music.

People are affected by color in different ways. Turquoise, for example, is a healing color for me; pink is the color of love. (If I'm wearing them both, watch out!) Ancient practices and recent research show how colors influence emotional moods. An artist friend of mine has decorated his dark home with pastel paintings; coming into his living room is like entering a garden. Another friend brings posters on business trips to hang on her hotel room walls.

The very tools with which we live and work affect us physically and emotionally. Computers have spawned an epidemic of tendinitis, carpal tunnel syndrome, eyestrain and back problems.

EXERCISE: THAT DEMANDING PHONE

Perhaps the most intrusive of all household instruments is the telephone. Are you aware of how your nervous system and muscles react to that ring? Do you always hold the phone with the same hand between your ear and shoulder?

To change your habitual relationship with your phone, you'll need to become aware of what you're doing. Here are some suggestions to guide you:

1. When you hear the first ring, take a deep breath and wait. Place your feet on the ground for balance.

2. As you hear the second ring, send your head and neck backward and sit on your sitz bones.

3. Reach for the telephone with your hand, leaving your body sitting back in the chair.

4. Bring the receiver to your ear without tilting your head.

5. Hold the receiver as loosely as possible.

6. After a while, hold the receiver in your other hand and change to the other ear.

Vocal Self-Care

Your voice is as unique to you as your fingerprints. Many people take their voice for granted until they lose it. But your voice is precious and needs attention and care. It's easy to strain your vocal cords by overusing them under stress—shouting, talking with a tight head or neck. My voice teacher had me chew a cracker and make humming sounds. "You can't strain your voice while chewing," he told me. Many singers warm up their vocal cord muscles this way.

EXERCISE: VOICE WARM-UP (SOLO)

1. Imagine you're chewing a cracker (or get a real one). Move your jaw up, down and side to side.

2. While chewing, begin to hum. Then open your lips and create a sustained tone. Explore your tone, from the lowest pitch to the highest. In the middle somewhere is your natural speaking voice.

3. Move your jaw up and down. Place your fingers gently in front of your ears on the TMJ joint, which connects your skull with your jaw. Chew, hum and sing, and listen to the movement your jaw makes.

4. Move your hands away, shake them out and rest. Then brush off any lingering tension from your face.

Emotional Self-Care: The Nourisher Who Gives Too Much

Being an empathetic "nourisher" can be rewarding, unless you nourish beyond your boundaries and comfort zone. People who are habitual nourishers wonder why they're continually drained and resentful after giving so much.

Marcie is such a nourisher. She adores her husband and her two young children, has a busy career and juggles a host of obligations. She is a woman to whom friends naturally turn when they're in trouble, for despite the demands of career and family, she gives tirelessly of herself. She cares.

Marcie has a niece, Jennie, whose life is in perpetual crisis. A runaway as an adolescent, Jennie took too many drugs, married too young, had three children before she was twenty-four, was deserted by her husband and now lives as a single mom in a poor section of town. Her mother is dead, her father has disowned her; there is no one to whom she can turn—except Marcie. Jennie calls Marcie every week, relates tales of woe and often asks not only for advice but for money.

Marcie is in a pickle. She doesn't really like her niece, resents her whining and perpetual requests for financial aid and would like to say no. At the close of each phone call, Marcie is drained, her teeth are clenched, her neck is tight and she's likely to feel nauseous. Clearly, her relationship to Jennie is toxic.

She resolves not to help. Her strategy is to dodge the phone calls, tell Jennie to look after herself, announce that there will be no more checks. She will practice tough love. But Marcie is in a double bind. No matter what she does, it doesn't work out for her.

If Marcie stops helping her niece she feels guilty—poor Jennie, her dead sister's only child—and lists a litany of concerns. It's not Jennie's fault she can't make a go of it; if Marcie won't help her, who will? What will happen to Jennie's innocent kids if Marcie doesn't help? So the phone gets answered, the advice is given, the check is sent. Marcie continues to feel resentful, to ignore her body's truth, and suffers. Marcie stews.

She is the prototypical woman who gives too much, denying herself emotional self-care. She overempathizes with her niece and has lost her own boundaries. These defenses/resistances saved Marcie when she was little, for she was made to take care of a vulnerable older sister and was only praised when she did so; but she's still in her family trance—always the caregiver.

Marcie is a close friend. I've watched her caught between her mother and sister—a dysfunctional triangle. Feeling she *had* to love her sister because no one else did, she continues as an adult to take on her family's problems.

"Who takes care of you?" I asked.

The question seemed to stump her. "Well, nobody really. My husband Jack does from time to time, but mostly I take care of myself."

"What would happen if you really didn't speak to Jennie? Didn't send her money?"

Another touchy question. "I'm not sure," Marcie finally said, then added quickly, "I'm afraid she wouldn't be able to pay her bills, she'd get sick and no one in the world would be on her side."

We were both quiet. I empathized with Marcie's dilemma. My family trance had the same theme, with different variations.

"And how would Jennie feel about you?"

"She wouldn't love me." Another long pause.

"What about your mother? Would she love you?" I asked gently.

"No. She'd disapprove and not love me," Marcie replied. "She rarely showed me affection or approval unless I did what she wanted." A sob caught in her throat as she realized the implications of what she was sharing with me. This was not a Rubenfeld Synergy session, so I ceased pursuing. She squeezed my hand and we changed the subject.

Marcie gives love generously to others the way she would have liked to receive it from her mother. She still caretakes Jennie, from her deep habitual need to be loved by everyone.

I've heard many helpers sing this same song with their own variations. "What do I need to do to change and how can I be more aware?" they ask.

Here's a partial list of suggestions. You might want to add more of your own:

1. Learn about your emotional boundaries and the consequences of overextending them.
2. Sense and listen to the early emotional and physical warnings when you give too much. They are truthful signals.
3. Trust your intuition (physical and emotional) when you engage with toxic people.
4. Practice saying no with clarity when you don't want to help or share your space or time. If guilt gallops in, say "Whoa" and deal with it later.
5. Be on the alert for people, particularly family members, who aren't satisfied until you agree to do what they want *all* the time.
6. Recognize other "nourishers" in your life—family and friends—and let *them* nourish *you!*
7. Be aware of your emotional stress states and ask for help from friends and professionals.
8. Quit the "Joan of Arc Club"—believing that no one else but you can save the world.

9. Become aware of your inner resources and strength. You're not "weak" if you admit that you don't know something or that you simply don't want to take on another responsibility right now. Recognize experts in other domains of life and use them.
10. Learn the difference between reasonable and unreasonable demands on yourself and others by experiencing the emotional fallout when you give too much and paying attention to it.
11. Review and practice the "Self-Care Toolbox" on pages 165–167. The tools work if you use them.
12. If your family trance screams, "You're going to die if you don't help," step aside and ask yourself, "Is that really true?" Your rational brain will explain that you have boundaries and that you will survive. However, your body and soul also need to believe that your survival is not predicated on accommodating everyone who asks for help.

"Bracketing"

The ability to empathize, feel and share in people's stories deepens human relationships. Clients and others share their experiences of betrayal, abandonment, abuse, love, agony and ecstasy. These themes may be awakened within you while you're listening. How can you continue to listen to the feelings of another when they bring up even stronger feelings of your own? How can you create some insulation between yourself and others without shutting them out altogether?

"Bracketing" is an essential skill for listeners who are helpers, healers and openhearted. As your own emotions begin to bubble up, they demand attention. Tell your anxious self to *wait* until after the session or meeting. Your worried self needs to hear that you won't ignore it, but that now is not the appropriate time.

In my early professional career, I had difficulty listening to Holocaust survivors. Because of my personal grief over the loss of many relatives, I experienced pain with each survivor. I dealt with this theme for years in my own therapy. Slowly I became able to be fully present as I listened, internally acknowledge my reactions as they arose, and then bracket them. As the therapist M. Lattanzi wrote, "I can be present by being close enough to the fire to empathetically feel the heat, yet separate enough to not be singed or need to flee."

When Helpers Don't Help

The helper's urge to help is not necessarily helpful. By rushing in [illegible] distress, helpers deny people the opportunity to discover what their pain is about. These people lose a chance to experience their innate capacity to heal and change. The message given to sufferers is: I know how to change you better than you do. I'm the expert.

Furthermore, people (and clients) become dependent when you don't give them a chance and enough time to help themselves. Helpers need to stand back and let their family members, friends or clients work out their own process.

"I *have* to help you," says the caregiver. The unspoken and unrecognized feeling is "I will be inadequate as a therapist, parent or friend." The reason the listening touch is so powerful is that it conveys total presence and empathy with minimal interference. When your ego is not involved in whether people accept your help, you will not get emotionally and energetically drained.

I offer you this gift of a "toolbox" so that you can come to the center of your own life with love and self-care.

A Self-Care Toolbox

1. *Humor.* Tell funny stories from your own life. Laugh at yourself. Invite friends to a "joke-athon." Dare to be outrageous. Watch funny movies and videos.
2. *Movement.* Dance. Wave your arms. Conduct your favorite music (unless you're sitting next to me at Carnegie Hall). Walk around your city neighborhood and in the country. Take exercise breaks at work. Engage in sports.
3. *Play.* Find games to play with friends. Allow yourself to be silly. Reconnect with the impish child you once were.
4. *Pleasure and Nourishment.* Create and participate in activities for enjoyment only. Receive massages, attend concerts, cook favorite meals. Be with loving, energizing friends.
5. *Feeding the spirit.* Find time to meditate. Be still or silent. Take solitary walks. Listen to inspirational music. Go to art galleries. Visit

working farms and pick your own fruits and vegetables. Grow plants, spices and flowers in your home. Garden. Remember these words of the Sufi poet Jabal al-Din Rumi:

Stop the words now.
Open the window in the center of your chest,
And let the spirits fly in and out.

6. *Rest.* Respect your body's need for sleep. Lie on the grass and daydream. Own a comfortable bed with fluffy comforters and supportive pillows. Use sheets that feel good. Wear comfortable nightclothes.
7. *Travel and Spas.* Choose interesting places to visit. Organize your travel activities so you're not exhausted when you return. Spend time at cultural and performing festivals. Stay in one place for an extended time instead of rushing from city to city. Choose a spa that offers sports, cultural activities, a good exercise program, well-prepared foods.
8. *Conferences and Workshops.* Go to conferences and workshops on healing and spirituality. Many of them take place in beautiful country settings, such as the Esalen Institute in Big Sur, California, or the Omega Institute in Rhinebeck, New York.
9. *Posture and Body Balance.* Practice ways of standing, sitting and moving easefully while engaged in everyday activities. Befriend gravity by balancing your body and thinking "up." Support your upright position by lengthening your spine and relaxing your head and neck.
10. *Love and Friendships.* Cultivate friends like a gardenful of colorful flowers, fragrant and mutually nurturant. Open your heart chakra to the love of family and friends. Cherish time with them. Share your life with people who love, respect and support you.
11. *Volunteering.* Contribute your time, energy and space to causes and people you believe in and want to help. Find ways to support them that will have meaning for you. Support the healthy future of our planet.
12. *Space and Environment.* Listen to the relationship between your inner and outer space. Design a workplace that is comfortable and energizing. Create a home you love, no matter how small. Allow your inner life to guide you.

13. *Reading.* Return to the experience of living with the characters in novels. Expand your mind through biographies and works on history, philosophy, spirituality and science. Curl up in bed with a book.
14. *A Wish List.* Create a list of wishes. Look at this list each day and notice which activity jumps off the page and calls out to you. I just met a couple in North Carolina operating a goat farm and an inn. These were on their wish list for years, until they realized they had better fulfill their dreams before it was too late.

Use all or some of these tools each day. Learning to use the "Self-Care Toolbox" can make the difference between burnout and being able to care for others.

A Balancing Act

There *are* solutions for self-care problems. All involve and invite balance. If you live and/or work in a constricted place, spend time in the country and other open spaces. If sounds around you create bedlam, find time to be still in a quiet room and listen to soft music. If you work in a closed, dark space, make sure to install good lighting and decorate with bright colors. A colleague painted a window overlooking a bright meadow on his windowless office wall.

Above all, leave time for humor, laughter and *pleasure.* The problems of your world are serious enough. Most people devote their lives to work and making money, gaining prestige and power. But the real rewards are psychological and physical: spiritual friendships, exercise and rest, love and emotional well-being. Balancing work with fun, stress with comfort, seriousness with laughter is the mainstay of a healthy body and soul.

11.

Couples: Sex, Spirit and Love

In the early 1980s, a couple at a workshop asked me to do a Rubenfeld Synergy session with them. At first I hesitated. But then, remembering how I learned music—starting with one voice and expanding to two, three and more—I agreed. Two tables were placed in the middle of the room. The husband lay down on one, the wife on the other.

Moving back and forth from one to the other while touching and talking was a remarkable experience. I could feel their subtle inner reactions to each other's comments; my listening touch told me whether their bodies were congruent with what they were saying to each other. The duet became a trio: the couple and me. While they were having a verbal dialogue with each other, my hands were communicating with their unconscious minds. I was able to help them develop the next movement of their life symphony.

Since then, I've often worked with couples. Sometimes they lie on separate tables, sometimes they sit opposite or facing each other. And almost universally, one of the first themes that emerges is the illusion of romantic love.

Romantic Love

"Romantic love is the single greatest energy system in the Western psyche," writes the Jungian psychoanalyst Robert A. Johnson. "Romantic love is not just a form of love, it is a whole psychological package—a combination of beliefs, attitudes and expectations. For romantic love doesn't mean loving someone; it means being 'in love.' "

Romantic love is full of impossible expectations. No flesh-and-blood partner can match the inner "ideal" of a mate. As the songwriter Lorenz Hart put it, "Falling in love with love is falling for make-believe," but men and women are real, full of frailties and contradictions, and the best that can be said of romantic love is that it is a doorway to a more important love, based on the slow opening of the heart and soul to another.

The story of *Beauty and the Beast* endures because it explores these notions. Beauty, held prisoner by the Beast, begins to recognize the Beast's heart essence—his shyness, clumsiness, gruffness, gentleness: his humanity. To the fury of the townspeople (society), who cannot believe that so beautiful a woman can abide so ugly a creature as the Beast, she rejects the Hunter, whose skin-deep beauty covers a cruel nature. When the Beast is dying (significantly, from a knife wound to his heart chakra), she is able to say to him, "I love you," and he can at last open his own heart to love.

But the story returns us to romantic love. (That's why it's a fairy tale.) For the Beast had formerly been the Prince, and with the Beast's demise the Prince reappears, handsome as ever. We are now supposed to believe that heavenly Beauty and gorgeous Prince create the "perfect match" and live happily ever after. We go from what to me is a true marriage—between yin and yang, masculine and feminine, ugliness and beauty—to the myth that society wants and expects: romantic love. In real life, inculcated with the myth, it's no wonder so many couples feel disappointed and betrayed when romantic love fades and reality intrudes.

Falling in Love with Potential

"You have such potential. I knew it from the moment we met. But you refuse to change."

This is a complaint I get from many couples. They fall in love because they see what *could be* rather than *what is.* But assuming your partner will become your ideal blinds you to seeing the real person. I am usually not surprised to hear, "I really didn't listen to, see or know him/her before we married. I just thought I did."

When couples first fall in love they enter a "love trance" state that usually overlooks differences. Their feelings of union, sexual attraction and euphoria are intensely highlighted. And on the heels of these feelings comes the engagement, followed by marriage.

In fact, when people get married they are not marrying an individual. They are marrying their partner's whole family and that family's mores and preconceived notions of marriage. Indeed, both the family trance and society's trance are so strongly entrenched in the unconscious mind that they undermine the intimate relationship between the couple. This often accounts for the fact that one of the partners "can't seem to change" or "doesn't want to change." Couples react, fight and communicate through their old unconscious "family trance" states, which manifest themselves particularly strongly in nonverbal communication and sexual behavior. Slowly and stealthily, resentment, mistrust and animosity creep into the relationship. Dalma Heyn's book *Marriage Shock* explores how women change the very behaviors their husbands loved before they married in order to fit into a marriage trance imposed by our society, with its rigid set of rules on how a wife should and should not behave.

Susan and Phil: Fulfilling Our Own Promises

There was a striking example of the family trance in operation at a recent workshop. Susan and Phil, a Canadian couple in their early sixties, approached me on the first day, anxious for a joint session. I had met them years ago and remembered them as a happy couple, but now their marriage was facing serious difficulties. He, a pessimist, was shy and withdrawn, unable to express his feelings. She was impatient and angry with him, frustrated by his passivity. They had grown more and more antagonistic toward each other. She wanted adventure; he wanted to stay at home. They were stuck, unable to resolve their differences harmoniously.

I set up three chairs and invited them to sit at my sides, facing each other. Phil's left shoulder felt like a wet mop when I touched it. "I'm so excited to be here," he said in a monotone which belied his words. He looked at his wife belligerently. "Susan's very optimistic. She believes thirty people will show up if we invite ten. But I know only five will show."

I touched his rounded back, and he closed his eyes. "What happened to little Phil when he was excited?" I asked.

"I don't remember," he said. "Maybe little Phil became excited when he played football." His left shoulder began to develop definition.

I turned to Susan and touched her right shoulder and upper chest. Her heart was beating quickly. I knew that she was barely able to contain her excitement, but she kept silent, giving Phil his turn. I asked them about their family backgrounds, a question I often use with couples. I want to get their position in their family constellation before I see how they interact with each other.

I returned to Phil's left shoulder. "I'm the fifth child of seven," Phil said. "When I was two, one of my brothers died. Now there are only three of us left. We were a working-class family and always struggled."

"Please describe six characteristics of your mother."

"She was very vocal in the community, critical, perfectionist, good cook, obsessively clean, sacrificed herself for people. She died four years ago," he rattled off flatly, with no emotion.

"Now describe six characteristics of your father."

"Withdrawn, prudent, workaholic, distant, noncommunicative, paranoid—a terrible fear of violence. He died six years ago." He spoke more haltingly and emotionally. I slowly moved my hands away from him as I turned to Susan, who had been listening attentively.

"Tell me a bit about your family, Susan."

"We were two, my brother and I, growing up in an upper-middle-class family." She shrugged. "A child of privilege. Lots of nice clothes, a big house."

"What are six characteristics of your mother?"

"Gentle, very communicative, sharing and an enabler—she made excuses for everyone—generous and kind. She died twelve years ago." Her eyes teared.

"And your father?" I asked gently.

"Alcoholic, workaholic, sometimes fun-loving and generous, kind—when sober—and controlling. He died thirty-three years ago."

"Phil, when you hear about Susan's family, what do you notice?"

"The obvious," he said, sitting up a bit. "Susan comes from a small family with money—they could do many things. I came from a big family, no money, always fighting."

"What attracted you to Susan?"

"She was so blond and beautiful, very different from my plain mom, who had no time to take care of herself or me." Phil began to whimper; I could see him regress. I placed my palms gently on his upper back. A painfully sad story unfolded about a little boy who spent three months in the hospital with a severe mastoid infection in his right ear. He described crying endlessly, and no one responding. He almost died. His mother could not come to the hospital very often, and now, as an adult, he could not share very much of his pain. "What was the use then and what's the use now?" he lamented.

He proudly told me he had regained three-quarters of his hearing. My listening touch sensed a major body shift. His rounded shoulders began to slowly melt from their frozen, painful past.

"Are you willing to do an experiment?" I asked, leaning toward his right ear. "You may not remember in your thinking mind, but your body remembers everything about those feelings of giving up. It's so wonderful that we can go back to our family trance-formations and change them." My left palm was on his upper back and my right fingertips touched his chest, directly over his heart, as though they were cradling it. He began to shake. "This little baby wants to be consoled, rocked and stroked," I said softly. "What would you have liked when you were in so much pain?"

He began to sob. "To be held!"

My hands continued to touch his heart area. "Whom do you want to hold you?"

"My mother. I felt her love through her hands," he said, holding his face, his eyes still closed.

"Imagine your mom coming over to your crib, picking you up and holding you. What does she say in a soft, loving voice?"

" 'It's okay, my little one. I'm staying with you and I won't leave you.' " His body softens.

I repeated these words to him. It was apparent to me that he feared Susan might leave him just as his mother had.

I saw Susan quivering, and I touched her lower back as I told Phil that he might try to imagine his dad helping out. Her quivering calmed, while Phil's shoulders moved in short, staccato spasms.

"You don't have to be silent," I said to Phil. "You can show your feelings and express your emotions. You *will* be comforted!"

"This baby's worth being heard!" he proclaimed. My hands were gently on his stomach and middle back. A deep sound began to rumble from within his chest. "I *am* worth being heard and comforted. Every part of me is being comforted right now. That's all that matters." He straightened, slowly uncoiling his spine, seemingly growing taller before my eyes. Every cell in his body responded to "I am worth being heard." An early "not worthy" trance-formation was changing to a "worth being heard" trance-reformation. His muscle texture felt lighter as he moved his shoulders around, letting go of a heavy burden.

"Phil, please open your eyes and look at Susan. Notice what you experience now."

He breathed deeply and smiled lovingly. "You do hear me, Susan," he said. "I'm worth being heard."

I suggested adding the words, "And you're not my mother."

He said it: "And you're not my mother."

Off in her own world, Susan began to tremble. She and Phil looked at each other. His words had brought her back to her own memories. "With an alcoholic father, there wasn't much I could do as a child," she said softly. After a few moments, her volume increased. "Mom always covered up, saying it was okay," she screamed. "Damn it, it's not okay! No one can make it okay!" She began hitting her knees. I quickly laid a pillow on her lap. Susan continued hitting and screaming—it was a full-fledged temper tantrum. I moved away to give her space. There was energy coming out of her head. We were witnessing how an early wounding could still gush with pain when a present-day action brought back a hurtful family dynamic.

"You're right. There's nothing your mom could have done to make your dad's drinking okay. Nothing! So what did *this* little girl decide early on?"

"To control my disappointment and never show it."

"What did Susan want to do very badly?" I asked.

"Go to the Ice Capades," she answered quickly. "But Dad didn't come home with the tickets and Mom didn't know how to drive." Her closed eyes were fluttering, a sign she was entering a deeper trance state.

"Oh, we can change the whole scene," I said brightly. "Imagine your mother waving the tickets in the air and saying, 'Guess who's got the tickets now.' Does she order a taxi?"

"No. A large, white, swanky luxurious limo. The kind with mirrored windows that no one can see inside. And seat warmers." If a back could smile, Susan's would have been grinning from side to side. "There's no alcohol, only a juice bar," she continued. "My brother stays in the limo because he loves all the gadgets. We're all dressed up. Mom's not excusing Dad for not showing up."

"How would you like your dad to react when you all came home?"

"He's happy we did this. He really notices that something different has happened with his family." Her eyes opened and sparkled.

"Bravo for finding a way to be creative, for going where *you* need to go, for going on *your* journey," I told her. "You're now making up for those past disappointments."

"If Phil wants to come along, that's great. But I'm going on my journey anyway."

I noticed that the energy around Susan had changed. She was flushed, her eyes were beaming and she was full of energy. Her agitation was gone; now she was alive with conviction.

"The more I feel comforted and take care of myself, the more I feel excited," he said. I asked him to repeat it with more ardor. He did, loudly. He was certainly being heard; he no longer needed to mumble. I had been moving back and forth between Phil and Susan. Now I sat in my chair, facing them. Susan turned to Phil and said proudly:

"We're fulfilling our own promises so we won't disappoint ourselves. We're changing our families' traditions, and waking up!" They got out of their chairs, kissed and embraced. Phil said, "Let's go!"

We all laughed, cried, cheered and danced. What else was there to do? This couple had moved from alienation to compassion, vitality and excitement. She no longer saw him as her father, he no longer looked at her as his mother. By changing their *individual* family trances, they could now make a commitment to working through their *mutual* family trance.

EXERCISE: CHARACTERISTICS OF YOUR PARENTS (SOLO)

Take this opportunity to look at the characteristics of your parents. Turn to your journal or a sheet of paper and begin.

1. Write six characteristics of your mother.

2. Write six characteristics of your father.

3. When you finish, take another sheet of paper and divide it into two columns. In one column, list six characteristics of your mother. In the second column, list six characteristics of your partner that are similar to the six you wrote down for your mother.

4. Repeat Step 3, only this time use your father instead of your mother.

Questions to Guide You:

- What are the similarities between your parents and your partner?
- What are the differences between your parents and your partner?
- What do you dislike about your parents that you don't like about your partner?
- What do you yearn or wish for from your parents that you expect from your partner?
- Which themes touched you and what are their wisdoms?

From Habitual Listening to the Listening Touch

Contact between people is essential for health and love, but maintaining contact in love relationships is a challenge to all couples. The pair must face their own projections, unfulfilled past needs, yearnings, resistances and differences and still reach out to a partner. By using a listening touch, people can experience acceptance, understanding, compassion, healing and love.

The ability to listen and to hear each other is key to an ongoing intimate relationship.

Many couples miscommunicate. Words are used to hurt, defend, avoid listening and rationalize. By suspending the use of verbal language and introducing the listening touch, we bypass the rational, judgmental mind and interrupt the habitual ways couples speak and listen to each other. The listening touch allows couples to discover each other's inner essence and beauty.

Before you begin these nonverbal exercises, you need to mutually accept some ground rules:

The Couples Agreement

This agreement is essential for creating safety and trust during the touch/movement communication exercises. I suggest each of you read and vow to abide by the following points.

1. I agree not to use these exercises to violate my partner's trust. If my partner wants to stop for any reason, I will respect his/her wishes immediately and without question.
2. I agree to hold my partner's behavior, information and feedback in the strictest confidence. Should we agree to seek professional counseling or couples therapy, we will discuss this with each other and come to a mutually appropriate decision.
3. I agree to let my partner express emotions and not interrupt him/her with "why" questions.
4. I agree to send my negative judgers (critics) to a beautiful vacation spot (if you send them to a terrible place, they will come back) so that I may be present with my partner.
5. I agree not to talk during the touch/movement sections so that we can both experience another way of listening to each other.
6. I agree to set aside ample time afterward to process the experience with my partner.
7. I agree to "check out" what I imagined was happening with my partner after we complete the exercise.
8. I agree that fun and play may be as important for us as the "serious" issues.

EXERCISE: THE HAND/TOUCH CONVERSATION FOR COUPLES (DUET)

I have led this hand/touch conversation exercise for many years, and am always moved by how deeply people connect in a short time. Many couples have reported new insights about their relationship after their experience.

Arrange a time when both of you are available to do the entire exercise without being interrupted or disturbed. You'll need two comfortable, moveable chairs, placed so your hands can touch your partner's hands easily.

1. Sit opposite your partner and observe him/her. How does he/she look at this moment? What do you notice? What are you experiencing in your body as you look at your partner? Is your judgmental critic on vacation?

2. Close your eyes, take a deep breath and do not talk. You are beginning a nonverbal journey with your partner using your hands and movement to talk. Rub your hands, shake them out and position them so that your palms face your partner's palms.

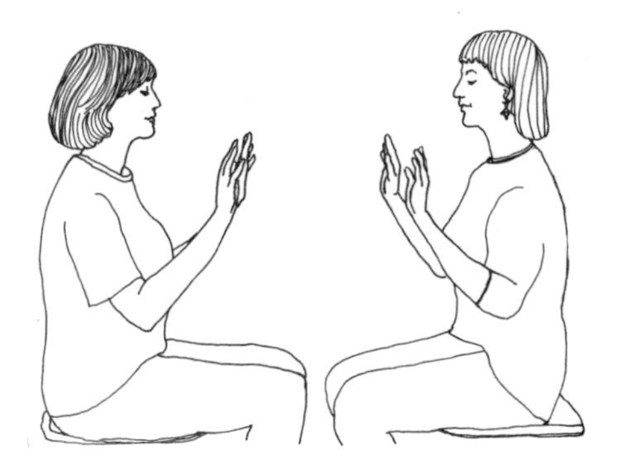

3. Move your hands toward each other, noticing any changes in the air and energy. When you make contact, take a deep breath and allow your palms to touch for a few moments in stillness. From here on, you'll go from Step 4 to Step 11 without opening your eyes.

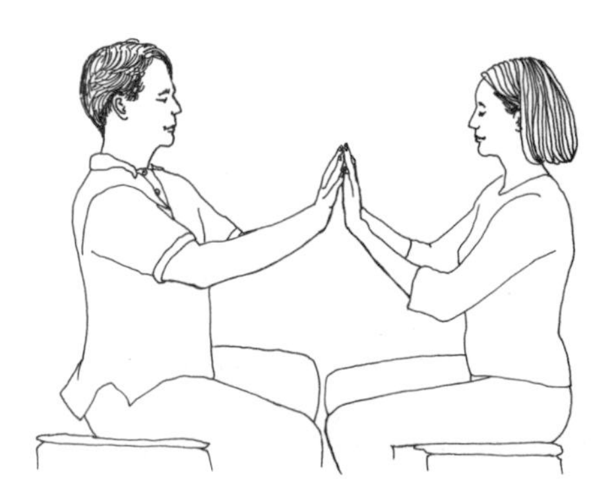

4. *Fun, joy and playfulness:* Imagine times in your relationship when you had fun together, when you played and laughed. Allow your hands to move and touch your partner's hands with fun, joy and playfulness. After some time, stop, separate your hands, shake them out and return them to your partner's palms. Take a deep breath.

5. *Anger:* Imagine a time when you were angry with your partner. Allow your hands to move and touch your partner's with anger. (Remember your agreement not to hurt each other.) After some time, stop, separate your hands, shake them out and return them to your partner's palms in stillness. Take a deep breath.

6. *Sadness:* Imagine a time when you were sad; you may not have expressed it at the time, and now you can. Allow your hands to move and touch your partner's hands with sadness. After some time, stop, separate your hands, shake them out and return them to your partner's palms. Take a deep breath.

7. *Curiosity:* Imagine a time when you were curious about each other. Allow your hands to move and touch your partner's hands with curiosity. After some time, stop, separate your hands, shake them out and return them to your partner's palms. Take a deep breath.

8. *Sex:* Imagine times when you felt sexual about yourself and your partner. Allow your hands to move and touch your partner's hands with sexy energy. After some time, stop, separate your hands, shake them out and return them to your partner's palms. Take a deep breath.

9. *Care and compassion:* Imagine times when you were filled with caring and compassion for your partner. Allow your hands to move and touch your partner's hands with care and compassion. After some time, stop, separate your hands, shake them out and return them to your partner's palms. Take a deep breath.

10. *Heartful love:* Imagine times when you opened your heart and experienced love for and from your partner. Allow your hands to move and touch your partner's with heartful love. After some time, stop, separate your hands, shake them out and return them to your partner's palms. Take a deep breath.

11. *Saying good-bye:* Continue your hand/touch conversation and say anything you wish. (Your eyes are still closed.) After a while, begin to say "good-bye" to each other. Take as much time as you need. Slowly move your hands away from your partner, shake them out and return them to your own space. Take a few deep breaths.

How couples say good-bye and leave says much about their temperament and timing. One partner says, "I want to say good-bye to every tree, the pool and the ocean." The other announces, "We're finished here. Let's hurry up and go." Both are fine and should be acknowledged without resentment.

12. Travel inside your body and notice any places that are asking for attention. Acknowledge them. Slowly open your soft eyes and gaze at your partner. After a few moments, share anything you wish.

Questions to Guide You Both:

- How does your partner look now?
- Were you able to convey your emotions to your partner?
- Did you receive your partner's emotions?
- Were you surprised at some of the conversation?
- How did your body experience this way of listening and talking?
- Did you get to know another aspect of your partner?
- What stood out for you in the hand/touch conversation?
- Were there any metaphors about your relationship?

Now take a sheet of paper and complete these sentence stems:

When I touched my partner's hands,
I experienced . . .
I wished . . .
I received . . .
I would have liked my partner to . . .
I resented . . .
I appreciated . . .
My body . . .
I am now aware of . . .

When I was playful,
I experienced . . .
I wished . . .
I received . . .
I would have liked my partner to . . .
I resented . . .
I appreciated . . .
My body . . .
I am now aware of . . .

When I was angry,
I experienced . . .
I wished . . .
I received . . .
I would have liked my partner to . . .
I resented . . .
I appreciated . . .
My body . . .
I am now aware of . . .

When I was sad,
I experienced . . .
I wished . . .
I received . . .
I would have liked my partner to . . .
I resented . . .
I appreciated . . .
My body . . .
I am now aware of . . .

When I was curious,
I experienced . . .
I wished . . .
I received . . .
I would have liked my partner to . . .
I resented . . .
I appreciated . . .
My body . . .
I am now aware of . . .

When I was compassionate and caring,
I experienced . . .
I wished . . .
I received . . .
I would have liked my partner to . . .
I resented . . .
I appreciated . . .
My body . . .
I am now aware of . . .

When I was sexy,
I experienced . . .
I wished . . .
I received . . .
I would have liked my partner to . . .
I resented . . .
I appreciated . . .
My body . . .
I am now aware of . . .

When I was full of heartful love,
I experienced . . .
I wished . . .
I received . . .
I would have liked my partner to . . .
I resented . . .
I appreciated . . .
My body . . .
I am now aware of . . .

When we said good-bye,
I experienced . . .
I wished . . .
I received . . .
I would have liked my partner to . . .
I resented . . .
My body . . .
I am now aware of . . .

Once you have finished the exercise, answered the questions and written in your journal, it is time for a different kind of exercise.

Centering and Grounding

Couples often find themselves pushed and pulled in all directions: by work, home, children, parents—and by the needs of their partner. Remaining balanced, centered and grounded is vital to your emotional and physical health. The following short exercise (adapted from the martial art aikido) demonstrates how thinking and imaging can change your body state and energy in a split second. If you feel yourself becoming emotionally and physically drained, practice the following exercise.

EXERCISE: WINGS AND ROOTS (DUET)

One of you will be A, the other B.

1. A, stand with your feet slightly apart. Close your eyes and image a cloud (or anything light) floating gently up in the air. All the energy in your body and mind are moving in an upward direction, like a bird spreading its wings and flying.

2. B, stand three steps away, facing A's back. While A images the lightness and upward sweep of a bird, walk over slowly and place your arms around A's waist. Take a deep breath and begin to move A's body upward. Notice the quality of A's body as you lift. Now gently let A come down. Move your arms away, step back and take a few deep breaths. Both A and B, shake your bodies out and take a deep breath.

3. A, again close your eyes and imagine you are a tree, sending your roots deep down into the earth. Allow your feet to sink into the floor and continue to send your "tree" energy down your entire body, through your feet and into the ground.

4. B, walk over to A, place your arms around A's waist and lift A upward. Notice what happens with A's body weight *this* time. Return to your original position and move your arms away from your partner. Both A and B, shake your bodies out and take a few deep breaths. Share anything you want about this experience. Now reverse roles and do Steps 1–4.

Questions to Guide You:

- B, how did you experience your partner's body weight in the first lift? Did you sense energy moving upward, like a bird flying high in the clouds?
- A, how did you experience yourself as your partner lifted you while you were imaging "up"?
- B, what happened when you lifted your partner the second time, when A was imaging the tree and root?
- A, how did you experience yourself as your partner lifted you while you were imaging "down"?

Sex, Spirit and Heart-Love

When couples first experience being "in love," it generally stems from physical attraction. That attraction often radiates from their sexual energy into deeper feelings. As the "courting" stage winds down, couples realize that their sexual life may not be the primary energy that binds their relationship. When they discover how to connect their sexual energy with heart-love, their relationship evolves and matures.

Although sexual feelings are spread throughout your body, your main sexual organs are located in your pelvis, your body's energetic traffic circle. Your spine is its "superhighway," providing the route for physical, emotional and energetic releases. It connects with your chest, neck and head. Side roads veer off, running to your shoulders.

However, sometimes sexual energy is blocked because there is no flexibility in the pelvic area.

Freeing the Pelvis

A young man was complaining about not feeling much in his pelvis. He confided that his attraction to women fizzled after he had been to bed with them a few times. I asked him to stand and move his pelvis forward and backward. He proceeded to move his shoulders, while his pelvis remained still. He *thought* he was moving his pelvis, but in reality it was frozen. No wonder he didn't experience pelvic sensations in his sexual encounters, I thought. I placed my hands on his shoulders so they wouldn't move and instructed him to move his pelvis again. What a struggle! Those shoulders were all mixed up. They *thought* they were his *pelvis.* Shocked and embarrassed, he agreed that what he imagined and reality were far apart. I asked him to look into a full-length mirror and move his pelvis forward and backward, while I continued to gently hold his shoulders. Slowly and steadily, he learned to send messages down to his pelvis to move. There was a joyous look on his face when he experienced his pelvis rocking for the first time. I let his shoulders go, and he moved around the room like an African dancer.

The pelvis houses the first chakra (sex and survival) and the second chakra (power). A free, open and energetic pelvis is essential for sensual and sexual pleasure. Here is an exercise that will open and free your pelvis.

EXERCISE: THE PELVIC CLOCK (SOLO)

I invite you both to do the Pelvic Clock as a solo first. Then we will go to the duet, which you will do with your partner.

Stand with your feet wide apart and your knees slightly bent.

1. Imagine a large circular clock face on the floor underneath you. (You are standing in the center of it, facing noon.) Place your right palm on your belly, slightly below your navel, and the back of your left hand on your lower back. Take a deep breath.

2. Allow your pelvis to tilt upward toward 12 o'clock (this moves your belly and lower back backward and your pubic bone up). Then tilt your pelvis downward toward 6 o'clock (this moves your belly and lower back forward and your pubic bone back). Remind your shoulders to be still for now; you want to isolate your pelvic movements first before they join in. Repeat this pelvic movement to 12 o'clock and then 6 o'clock several times. Exaggerate the extremes of forward and back, up and down, and take a few deep breaths. Allow your hands to continue touching your belly and lower back; this will remind you to focus on the pelvic movement. Notice what happens to your lower spine as you move your pelvis this way.

3. Now place your hands on your hips. Sway your pelvis toward 9 o'clock (to the left) and 3 o'clock (to the right). Practice and exaggerate these hip movements several times. After several repetitions, shake your hands out and take a deep breath.

4. Place your hands on your belly and lower back. Again, tilt your pelvis toward 12 o'clock, then 6 o'clock, to 9 o'clock and to 3 o'clock. You're creating a line from 12 to 6, and a line from 9 to 3. Repeat these two directions several times. After a while, move your hands away, shake them out and take a deep breath.

5. Stand with your knees bent, and return your hands to your belly and lower back. Tilt your pelvis toward 12 o'clock, then move it slowly to 1, 2, 3, 4, 5, 6, 7, 8, 9, 10, 11 and finally back to 12 o'clock. You have just moved your pelvis in a complete clockwise circle. Repeat this circular movement several times. Now reverse your circle. Tilt your pelvis toward 12 o'clock, then move it slowly to 11, 10, 9, 8, 7, 6, 5, 4, 3, 2, 1 and back to 12 o'clock. You've just moved your pelvis in a complete counterclockwise circle. Repeat these circular movements several times.

6. Move your hands away from your pelvis and hips and shake them out. Repeat the pelvic clock movement with your hands floating around anywhere you wish. If you have some Caribbean music handy, play it. Move freely around the room, letting your entire body experience pleasure and joy.

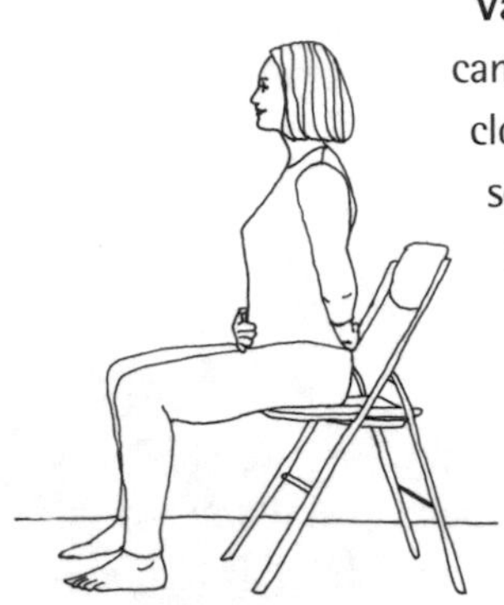

Variation A (useful in an airplane or at the office): You can do the Pelvic Clock sitting in a chair. The image of the clock will still be below you: on the floor or on the chair seat. Your pelvic movements will follow the same directions as for the standing exercise. Notice if your physical sensations are different.

Variation B (lying down): You can do the Pelvic Clock lying down with your arms resting at your sides.

1. Float your knees up toward the ceiling and place your feet on the floor about two feet apart.

2. Imagine the clock in front of you, instead of underneath.

3. Move your pelvis to 12 o'clock. Your lower back will flatten onto the floor and your pubic bone will tilt up.

4. Move your pelvis to 6 o'clock. Your lower back will arch, and your pubic bone will tilt down.

5. Tilt your pelvis toward 9 o'clock (left). Your right buttock will lift slightly off the floor.

6. Tilt your pelvis toward 3 o'clock (right). Your left buttock will lift slightly off the floor.

7. Move your pelvis in a circle clockwise and then change the direction—counterclockwise. Your body sensations will differ because you are lying down.

8. When you complete these pelvic movements several times, stop and take a deep breath.

9. Roll over onto your side, wait, then slowly sit up. After a few moments, move up to standing. Notice any shifts in your body.

10. Share anything you wish with your partner or write in your journal.

The duet part of this exercise (and those to follow) will guide you both through a series of *touches* and *movements*. One partner will "listen" (using touch), while the other partner will "tell his/her story" (using movement). Each person will have an opportunity to be "the mover" and then "the listener." There are questions and suggestions at the end of the exercises to stimulate your discoveries and to support your change process. (In Chapter 13, there is a Seven-Day Program with additional exercises that you can do alone and together with your partner.)

EXERCISE: THE PELVIC CLOCK (DUET)

Preparation: One partner will be A, the "listener-toucher." The other will be B, the "mover-storyteller." You will each have a chance to play both roles. Please review the Couples Agreement on page 176 now.

1. B, stand with your feet wide apart and knees slightly bent. Close your eyes and imagine the clock on the floor underneath you. Travel inside your body and go to a time when you enjoyed moving your pelvis freely.

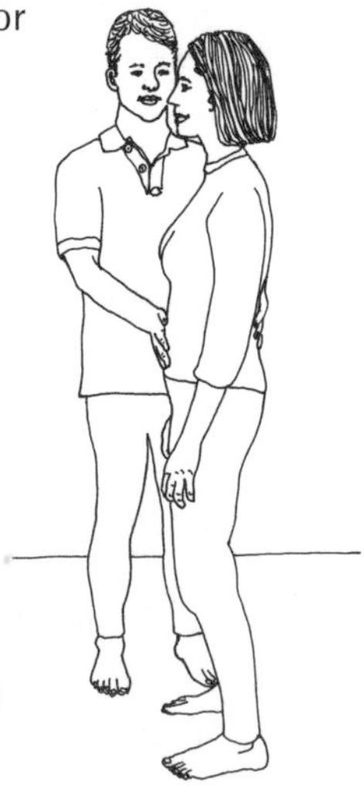

2. A, stand facing your partner's left shoulder, close enough to touch her/him with your hands. You should be able to look into B's left ear. Close your eyes and send your awareness to your hands. Thank them for being open and listening to your partner.

3. A, open your eyes, rub your hands together and shake them out. With a clear intention of "listening," slowly and gently place your right palm on your partner's belly (right below the navel) and your left palm on your partner's lower back. Now close your eyes again.

4. B, begin to tilt your pelvis forward toward 12 o'clock and then 6 o'clock (the same way you did the solo). Let your awareness move to your pelvis and your partner's hands.

5. A, gently follow your partner's movements with your hands. If you are tempted to "correct" or change your partner's movements, stop and return to the principles of listening touch. This is your partner's movement-story, not yours.

6. B, now tilt your pelvis from 9 o'clock to 3 o'clock (side to side) several times. Notice how you feel.

7. B, now move your pelvis in a clockwise circle. Then move it in a counterclockwise circle. Repeat these circular movements several times.

8. B, move your pelvis to 12, 6, 9 and 3 o'clock (in two intersecting straight lines). Now move it in a circle. Combine and alternate the circles with lines.

9. B, when you have completed several pelvic movements and want to stop, let yourself be still and stand quietly. Listen to your body and notice how you feel.

10. A, slowly move your hands away from your partner's pelvis and shake them out. If your eyes are still closed, open them slowly. Wait for a few moments, then move to face your partner. When your partner opens his/her eyes, you'll want to be there as a "loving greeter."

11. B, open your eyes slowly and gaze at your partner. Share whatever you wish with each other.

Questions to Guide You:

- Listener-toucher A: How are you experiencing listening to your partner's story with touch. What thoughts and feelings emerge from you? Are you able to let your partner travel on his/her own journey?

- Mover-storyteller B: How are you experiencing moving your pelvis while your partner touches/listens to you? Do you feel free to move and express your story?

12. Now reverse roles. A, the listener-toucher, will become the mover-storyteller. B will become the listener-toucher. Repeat Steps 1–11.

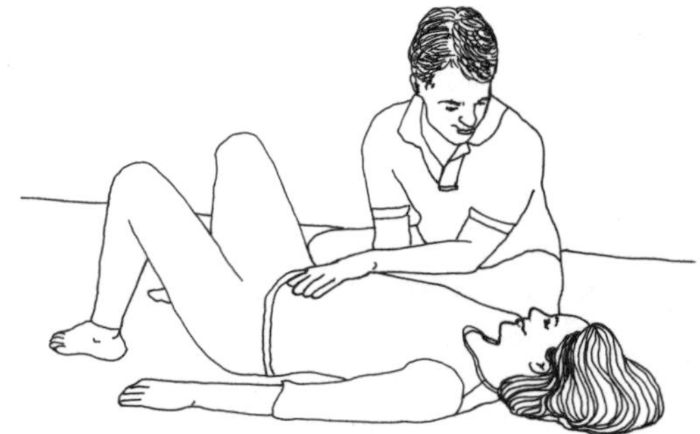

Variation A: You both may do Steps 1–11 once again with one partner lying down (knees floating up with feet on the floor) and the other touching with his/her listening hands.

Questions to Guide You:

- How do you see each other now?
- How did you experience being touched while you moved?
- How did you experience touching and listening while your partner moved?
- Did you experience being out of sync with your partner at any time? When?
- Did you experience being in sync with your partner? When?
- Did you gain some wisdom about your partner?
- Did you gain some wisdom about yourself?

Opening the Heart

The "superhighway" between pelvis and chest, neck and head needs to flow freely so you can experience the harmony of sexual pleasure and heart-love. An energetic traffic jam will congest the highway and impede the connections you need in order to feel whole and integrated. Since flexible shoulders and upper backs are essential to opening the heart area, we'll turn our attention to that area in the next exercise.

EXERCISE: CHEST AND SHOULDER OPERA (SOLO)

Locate a quiet and comfortable space. Set aside enough time to complete this exercise as *solo* and *duet.* You'll need a firm carpet or mat and a few pillows. Have your journal and writing utensils ready.

1. Lie down on your back, stretch your legs out and rest your arms at your sides. Close your eyes and gently roll your head from side to side. (Take a "memory photograph" of this rolling movement.)

2. Travel through your body and notice which areas make contact with the floor and which do not.

3. Send your awareness down to your knees, allow them to float up toward the ceiling and place your feet on the floor.

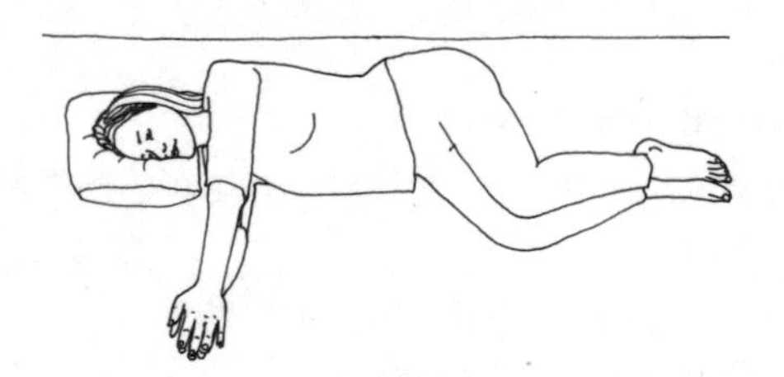

4. Roll over onto your right side. (Place a small pillow or rolled-up towel under the right side of your face.) Extend both arms straight ahead. Your right arm will rest on the floor (palm facing up to the ceiling). Your left arm will be on top of the right arm (left palm covering the right palm). We'll call this position "home base." Your legs will be bent at about a 90-degree angle, with your left leg on top of the right. Take a few deep breaths.

5. Imagine that you are a trombone. Your left arm is the slide, moving forward. The trombone slide is straight; it does not bend. Now, *slide* your left arm

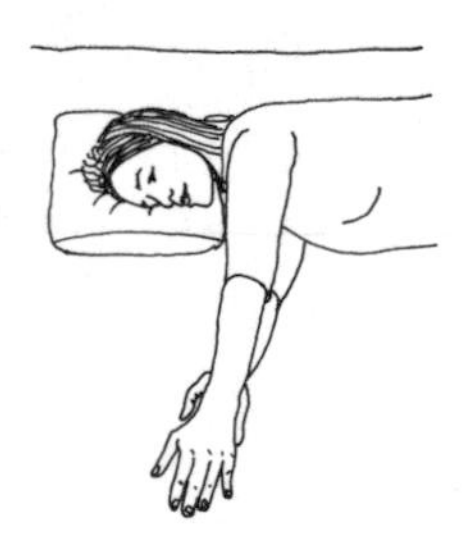

forward as though reaching for something. Your palm will touch the floor as it continues to move forward. (Remember, your elbow is straight and your left leg remains on top of the right one.) Now slide your palm and arm back to "home base" (left palm covering the right one). Repeat this movement several times.

6. *Slide* your left arm backward toward your chest (like the trombone slide) and then return to "home base." Repeat this movement several times.

7. Now slide your left arm way forward (reaching) and then backward toward your chest. Repeat this forward-backward movement several times. While your arm is moving, send your awareness to your chest and your left shoulder. What do you notice?

8. When your left arm moves easily, change the rhythm of the movement—sometimes slow and sometimes fast. Let your head and neck follow in any way they wish. After a while, stop, shake out your hands and rest.

9. Roll over onto your back and rest your arms at your sides. Float your knees up toward the ceiling and place your feet on the floor.

10. Roll your head from side to side. What are the differences now from the first time you moved your head? Take a deep breath and listen to the expansion and contraction of your chest and rib cage. Are there any differences between the two sides of your body? Create any images and metaphors that express these differences.

11. Your other side is jealous. Follow Steps 1–9, this time moving your *right* arm as the trombone slide.

When you have both completed the solo part of this exercise, share anything you wish with your partner.

EXERCISE: CHEST AND SHOULDER OPERA (DUET)

Review the Couples Agreement before you begin. Decide who will be A (the "listener-toucher") and who B (the "mover-storyteller"). You will reverse roles when you complete the exercise.

1. B, lie down on your right side. (Place a small pillow or rolled-up towel under the right side of your face.) Bend your knees at about a 90-degree angle, with your left leg on top of the right. Extend both arms straight ahead. Close your eyes and travel through your body. Imagine a time in your life when you reached out for someone or something.

2. A, place a couple of pillows under your sitz bones so you can cross your legs easefully. Sit on the floor facing your partner's back. Rub your hands together and shake them out. Send your awareness to your hands, then thank them for being open and listening to your partner.

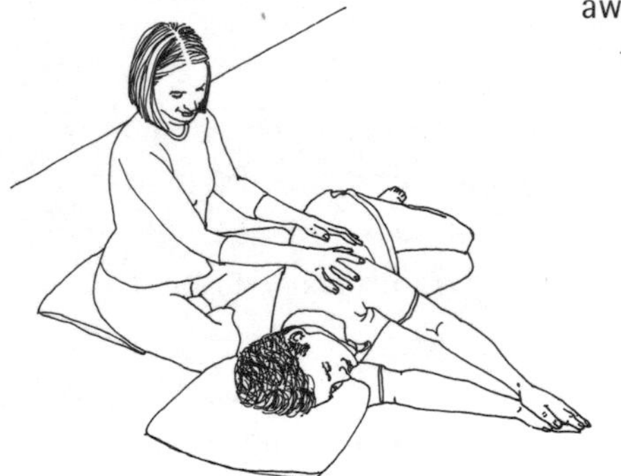

3. A, create an image (such as a butterfly) and *slowly* move your right hand so your touch will be very light. Gently place your palm on your partner's left shoulder blade. Move your left hand toward the rib cage and gently place it on your partner's side just above the waist. Using your listening hands, follow your partner's movement-story. The lighter the touch, the more you will hear your partner's story.

4. B, you've already practiced these movements as a solo, so they will be familiar. Slide your left arm forward like a trombone, keeping your elbow straight. Now, slide your left arm backward toward your chest, again keeping your elbow straight. Slide your arm forward again, as if you were reaching out for something. Repeat the forward and backward movements several times.

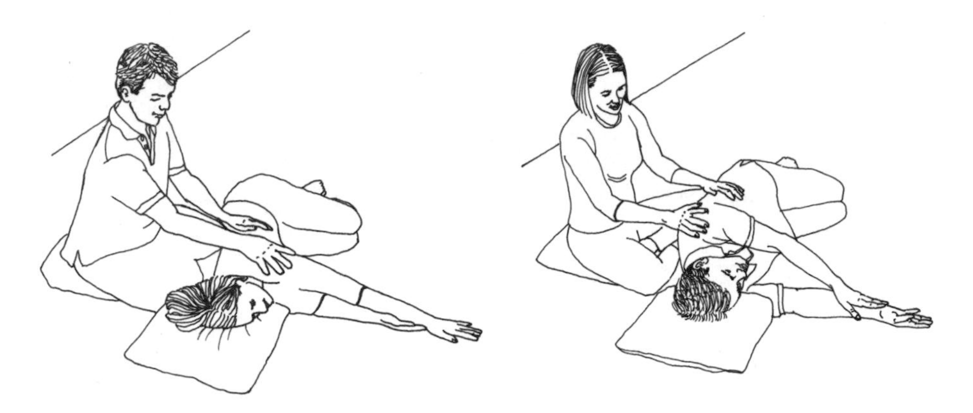

5. B, as you move, allow a story and images of people to emerge. Perhaps you're little and enjoying the sound of this trombone. Perhaps you're older and you recognize emotions in your shoulder, back and chest. As these scenes unfold, they may become metaphors for deep issues and behaviors in your life. Your partner will be following and listening to you.

6. A, allow your hands and your total being to listen and follow your partner's movements. (If the negative judgers return, shake your head and send them back on vacation!)

7. A, slowly move your hands away from your partner and shake them out. Rest them on your lap. Wait for your partner to complete his/her movement-story.

8. B, complete your forward and backward movements. Roll onto your back, stretch your legs out and rest your arms at your sides. Take a deep breath and slowly open your eyes. Your partner is waiting to greet you with a big smile.

9. Gaze into each other's eyes. When you both feel ready, share anything you wish about this experience.

Questions to Guide You Both:

A (The Listener-Toucher):

- How did you experience listening to your partner's life's movement as a metaphor?
- Are you willing to accept your partner's way of moving in life?

- Did you have an urge to "correct" your partner?
- If your partner seemed confused, did you want to straighten him/her out?
- Did you also have some images?
- Tell your partner how you experienced yourself while touching?

B (The Mover-Storyteller):

- How did you experience your partner's touch as you moved?
- Did you allow yourself to receive unconditional listening?
- If you experienced your partner pushing, did you tell him/her to stop?
- If you had any images while you were moving, please share them.
- Tell your partner how you experienced yourself while being touched.

For A and B:

- What do you notice about your heart area when you inhale and exhale?
- If you were to do this exercise again, what changes would you like?
- Did you experience a connection between your pelvis and heart?
- How do you see and experience each other at this moment?

Now, switch roles. A will be the mover-storyteller and B the listener-toucher. Follow Steps 1–9.

When you've completed your verbal exchange, slowly stand and walk around the room. Bravo!

Movement and Listening Touch

The struggles and joys in your relationship are powerfully told by the way you move and the way you touch your partner. Your life as a couple is happening right here and now. The moment is special. Your relationship has been enriched by listening to each other in this gentle and respectful way.

As a grand finale, here is a last lovely touch/energy exercise to further enrich your heart-love connection with your partner.

EXERCISE: THE HEART'S SONG (DUET)

Sit comfortably opposite your partner, either on the floor or on chairs. The two of you will be doing this exercise at the same time.

1. Rub your hands and shake them out.

2. Place your left palm gently over your belly, just below the navel, and gently touch your partner's heart area with your right hand.

3. Gaze into each other's eyes and take several deep breaths.

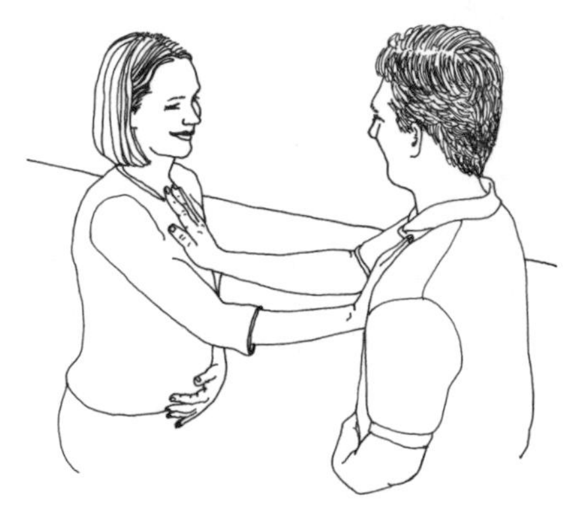

4. Now close your eyes and travel to a time when you felt accepted and experienced heartful love.

5. Open your soft eyes and gaze into each other's eyes again.

6. Slowly move your hands away and rest them in your lap.

7. Now switch hands. Place your right palm over your belly and gently touch your partner's heart area with your left hand.

8. Listen to each other's heart song.

You both deserve to play, have fun, experience pleasure and laugh with each other. When a situation becomes too serious, lighten and enlighten yourselves with humor.

12.

Humor and Other Martial Arts

Humor made my family's troubles easier to bear. When my father lost his job and we had no food, he and I went to the local Hungarian restaurant and "begged." When we got home, I pointed out the flaws of the food—how the stuffed cabbage was massacred, how the vegetables were overcooked. "I bet they were glad to get rid of this stuff," I told my parents cheerfully. Making fun of the food made the shame of begging bearable. I used the same technique later to cope with the physical pain and frustration of being a female conductor. I certainly used humor to reject my mentor Fritz Perls's sexual advances, inappropriate even for the late 1960s. Indeed, it was Fritz's "clown"—a persona he invented as part of his therapeutic techniques—who taught me the value of humor, demonstrating how close tears and sadness are to laughter.

My humor rescued me from my psyche's darkness and hopelessness; it covered up a great deal of pain as well. As my sense of humor blossomed with my teachers, it also naturally developed with my clients. I began to use it as a significant part of my work. To me, a sense of humor demonstrates an understanding of human suffering and misery and an acceptance of our human weaknesses. Laughter helps us detach from painful situations and emotions. It is a sign of choice and freedom.

Humor is all-pervasive in Rubenfeld Synergy, but only as an accepting, open, gently embracing attitude toward life. It is an important ingredient in many of our sessions; it can change a mood, illustrate the opposite of darkness, dissipate fear and anxiety and create the safe environment that is essential for people who dare to open their eyes to see all of themselves. For both Synergists and clients, humor is of great value for understanding, remembering, learning and enhancing the change process. It can be lifesaving.

The use of humor reframes stressful situations, diverts and distances clients from their profound pain (albeit momentarily) and instills a closer relationship with their therapists. And laughter releases physical and emotional tensions. In session after session, we witness how an accepting attitude, an openness to life's absurdity—as well as its sacredness—helps the client open up to new self-awarenesses and insights.

There is a difference between just telling jokes and using situational humor in a Rubenfeld Synergy session. I often open my workshops with a joke or two to give participants and myself an amusing way to settle in. Afterward, my sense of humor becomes integrated into the therapeutic process. In Rubenfeld Synergy, humor and laughter may evolve spontaneously or be purposefully used.

Many humorous moments arise from incongruity or paradox; we build on them with wordplay and exaggerations. During sessions, clients are not afraid to initiate their own funny stories and comments, because they feel safe.

You can always tell when clients are getting better, a therapist friend noted—they start to laugh at themselves. When events seem unbearable, humor cuts through and becomes the glue that keeps the psyche together. Humor in wartime, for instance, is legendary (which is why *M*A*S*H* was such a long-running hit). To find something to laugh about in the face of catastrophe helps us cope—whether the catastrophe be global or individual, physical or emotional. We can diffuse a potentially terrifying experience with a joke; a laugh can have as much emotional power as a sob. Funny stories and outrageous scenarios can open a door to wellness. They present a different perspective and interrupt "painful, hopeless and doomed" trance states.

Some people react well to humor, some are offended by it. In other words, its use is a *choice* that you make when you are confronted with ongoing stress or trauma. Although problems may continue, humor is a time of respite, and it changes the physical energy of the people who laugh.

Humor needs to be used appropriately if it is to have therapeutic benefits. A practitioner must rely on his/her wisdom, skill and judgment in deciding when to use it. As with listening touch, intentionality is vital. Sarcasm, contempt or mockery have nothing to do with healing humor. In Rubenfeld Synergy, we would not use humor to belittle, mock or deride clients. Negativity and sarcasm do not support trust. On the contrary, they deepen the wounds that are already present. Humor is primarily used to release fear and make way for the healing process. It is an exchange of love through smiles and laughter—a process that is essential for parent/infant bonding and thus for survival. (In later childhood, laughter can be associated with pain if we are laughed *at.*)

Humor and Healing

Laughter was a well-known remedy centuries ago. The Bible tells us, "A merry heart doeth good like a medicine," and according to Voltaire, "The Art of Medicine consists of amusing the patient while nature cures the disease." Many ancient Sufi, Zen and Jewish folktales teach wisdom with humor. The Fool and the Clown are archetypal and universal characters who live in all of us, and too often remain silent.

In the name of science, old wisdoms about the healing qualities of humor were lost along the way among health professionals in the Western world. But that same science is now showing us its validity. When Norman Cousins's book *Anatomy of an Illness,* an account of his recovery from a "fatal" illness through the use of laughter, became popular in the 1980s, a new appraisal of laughter's healing role was forced on the medical profession. There is abundant evidence that a smile and laughter have beneficial health effects. Mirthful belly laughter has been called "internal jogging" because it creates a workout for the whole body. And as with body, so with mind. As the homeopath Dr. Henry Williams puts it, "Humor is a lightness of touch with reality that helps me be more in touch with how I feel. It takes away the deadly seriousness and makes space for the healing to occur."

In sum,

- Humor provides ways to express unacceptable feelings by allowing you to talk about dark emotions, fears and shame in trusting, safe, nonthreatening ways.

- Humor shifts an individual's frame of mind so that he/she can o[illegible] and make contact with others.
- Humor supports communication about sensitive matters in a light way.
- Humor has the power to change tensions, create a relaxed atmosphere and facilitate healing.
- Humor can lead to insights about physical and emotional conflicts.

Let's see it in action.

The Art of Suffering

There are real reasons for suffering. But some cultures—and some individuals—make suffering into an art form. Often, there is only an imaginary reason to suffer, yet some people still respond to everyday events as if they're in a life/death crisis. "Oh, you're killing me," says the mother. "You'll be the ruin of this family," says the father. "You're breaking my heart," the girlfriend announces. "I'll die if you leave," the boyfriend pleads.

Suffering and guilt are great pals. I see them hanging out and holding hands at workshops.

A workshop participant began telling his tale of woe. His boss didn't appreciate him. His children were no good. His wife didn't love him anymore. He couldn't abide his mother. He went on endlessly, suffering with greater and greater intensity as the list got longer.

"Are you Jewish?" I asked.

"Well . . . yes."

"I thought so! Make soft fists, rock back and forth and hit your chest. Say, 'Oy vey, you're making me suffer. Oy vey.' "

I asked him to continue repeating this phase. Pretty soon he began to smile, and then he was laughing uproariously and so were the other workshop participants. He had been blind to himself as a lifetime "sufferer." Now, he was finally recognizing how he exaggerated and distorted his reactions to life. Humor and laughing interrupted his "suffering trance" state. Sometimes, however, the pain is so intense that the psyche says, "Enough, we're shutting down," and defensive curtains are drawn.

Laughter and the Dissipation of Fear

Fear leads to physical and behavioral paralysis. Humor unlocks the paralysis, enabling people to delve more deeply into their dysfunctional behaviors. Just as martial arts like judo, aikido and tai chi use the energy of opponents to deflect and defeat them, humor uses negative energy of the past and transforms it to a positive energy of the present. This is important in Rubenfeld Synergy, because fear tightens the body and humor helps melt it.

Peg, a student of mine, was reluctant to confront conflict and always seemed to hover in the background. She came to me one day and admitted her fear. "I can't even speak with you," she said with terrified eyes.

"Well then, let's bark," I answered.

She growled and yipped. We burst out laughing. Later, we were able to discuss her problem without constraint.

Humor and the Acknowledgment of Feeling

Millie, a vivacious doctor, complained that her life was too busy, that she was always "running around." "I want to slow down," she told me. "I'm too old for this sort of thing."

Yet when I touched her, her body was loose, energetic and pliable. It seemed to say that she was excited by her life, interested in everything. I asked her to get off the table and start moving around the room. I joined her, and soon both of us began running, circling the room again and again until we were out of breath. Finally we both stopped, looked at each other and burst out laughing.

"All right," she said. "I *love* running around. It makes me feel alive."

Exactly.

Marsha, who had been brought up by two extremely moralistic parents, proclaimed proudly that she was good at her job, always serious, punctual, disciplined and undemonstrative. Throughout her childhood, she had been told and reminded that sex was "dirty." When she had "impure" thoughts, she not only hid them from her parents—she hid them from herself.

"What price did you pay for hiding?" I asked.

"I never allowed myself to experience my sexuality unless I was drunk or stoned." She blushed. "Not that I got drunk or stoned that often."

"You felt guilty?"

"Yes."

"What are you experiencing now?"

"Um, well, there's sadness coming up."

"Sadness." I nodded. "And now?"

"Sadness," she said again, mumbling. "The loss of all those years . . . I don't know. I just . . ."

"What?"

"Feel sadness, I guess."

"And now you're forty-nine."

She grimaced. "Very much so."

"Yeah."

"And getting older by the minute."

"And getting older by the minute," I repeated, copying her doleful tone. "We better pack up our bags and *die now.*"

The group's laughter, as much as my words, shocked her into a realization of the humor in what she had said.

"Actually," she told me, her face flushing, "my secret desire is to be Carol Burnett!"

This time the group exploded with laughter. "Well, you are!" I exclaimed. "You're a riot, an absolute riot!"

She glared at me. "But that's ridiculous! I'm the most serious, hardworking, responsible . . . *guilty* person in the world!"

"Good for you," I told her, beating on my chest. "Oh. Come on! Responsible! Serious! Hardworking!"

She followed my example, and soon we were both chanting the words. "Responsible! Serious! Hardworking!"

"Don't forget guilty!" she shouted, pounding on her chest.

"Guilty! Guilty! Guilty!" I repeated with her, and soon the class was chanting along with us. I looked at Marsha. She was sitting on the edge of the table, tears of laughter streaming down her face. She became vibrant, *sexy.* She was ready for change.

Exaggeration

Exaggerating humorously highlights the contradictions between what we say, what we do and how we feel. I was reminded of these contradictions recently when Judy, a workshop participant, told the group that she was "trying to get a job."

"How many interviews have you had in the past month?" I asked.

"Well . . . none."

"And how many resumes did you send out?"

"I haven't sent them yet. They're being prepared. I'm *trying*," she whined.

"Ah, you know *all* about *trying*," I said. "Just sit there, don't move, and *try* to get a glass of water."

"What's the matter?" she replied huffily. "You don't believe in kinetic?"

"Sure I do," I answered with a smile. "Kinetic food. Kinetic money." We both laughed.

EXERCISE: "KEEP TRYING" (SOLO)

Imagine something you want. Sit very still while you mentally exaggerate trying hard to get it. For example, *try* and work hard at *letting go. Try* and work hard at *being free. Try* and work hard at *not worrying. Try* and work hard at *not trying.* Stop for a moment. What sensations and feelings do you have when you *try* hard and *don't move. Trying* is like being "a little pregnant."

Laughing for the (Physical) Health of It

Laughing not only helps release muscle tensions, it also improves blood circulation, increases the oxygenation of the blood, enhances digestion, reduces pain (because of the release of endorphins, the brain's own pleasure-producing energy) and best of all strengthens the immune system (via its effect on the neuroendocrine and stress hormones).

It has been shown that watching a funny video helps the lymphocytes proliferate and the natural killer cell activity increase. Even *anticipation* of an experience likely to induce laughter seems to reduce stress hormones and improve immune function.

The general pattern of the impact of laughter on the body is one of stimulation followed by relaxation, which we can easily notice in our muscles. Actually, laughter is probably the most powerful stress reliever we have, because it is almost impossible to maintain muscle tension and laugh at the same time. Even *faking* a smile or a laugh works. The very act of lifting the corners of the mouth—"mouth yoga"—affects our body chemistry and makes us feel better. Try it sometime as a tiny exercise. You'll be amazed by its effectiveness.

Sometimes I embarrass my friends when, overcome by laughter, I slap my knee and make loud sounds and snorts. I don't care—I just let loose. My breath sweeps all the toxins of sadness out of my body; my rib cage expands to let in new energy.

Laughter is very physical. Through my hands, I can feel the dramatic body shift from tightness to softness after one has laughed. Not only has humor interrupted the pain cycle, it physically changes the body areas holding the pain. In anger, energy is dense, almost congested. But with laughter, energy ripples and flows, building up momentum.

Adapting Humor

To use humor therapeutically yourself, you'll have to practice and take risks. Sometimes, to gauge the level of humor potential in the group, I'll start a workshop session with a funny story or a joke that has been successful in the past—such as the following:

A lady from Brooklyn calls up her travel agent and says, "I'm seventy-five years old, and I must have the answer. I *must* go to a monastery outside of Nepal."

"Why there?" the travel agent asks.

"I've got to see the guru!"

The agent is horrified that so frail a lady would undertake so arduous a trip. "It'll cost you five thousand dollars and you'll have to trek for five days," he tells her.

"I don't care. I've got to see that guru!"

He books her on the long flight. When she arrives in Nepal, she's greeted by a group of Sherpas, who reach for her luggage.

Coming from New York, she clutches her purse. "No one touches it but me," she exclaims.

They trek for five days through the wild and rugged countryside. Finally, she arrives at the monastery, clothes ripped, face and body scratched, exhausted—a wreck! A young Buddhist monk greets her.

"Sorry," he tells the lady, "The guru has vowed silence. No one can see him."

"Listen," she says strongly. "I'm from Brooklyn, I just spent five thousand dollars and I trekked for five days. You *have* to let me see him."

Reluctantly, the monk agrees. "Okay," he tells her, "but you have just ten seconds."

He leads her to a sacred room, filled with a hundred statues of the Buddha and the scent of incense. And there, at the far end of the room, sits the guru himself, dressed in saffron robes, his fingers touching, his eyes rolled upward, chanting, "Ommm."

The little lady marches right up to him, takes her purse and whacks him across the chest. His eyes open wide in shock. "Harold," she says, "I'm sick of this. Come home right now!"

And then I add, "Every guru has a mother."

I use the story to emphasize to my clients that all the important answers are within them. The best support that teachers, therapists and, yes, gurus, can give you is to teach you the tools to discover your own innate wisdom. Everyone you put on a pedestal also has to deal with life's mundane realities. Even therapists must take out the garbage.

Questions to Guide You:

- What were your associations and reactions to the guru joke?
- Can you imagine a teacher or mentor and consider that he/she may *not* have all the answers you seek?
- Where would this joke be appropriate?
- Which of your friends or clients need to hear it?

I'm going to end with descriptions of two sessions in which humor and high seriousness were combined.

Gregory:"By George, You've Got It!"

Gregory, a school psychologist, rushed to the table and proudly announced that he had worked with Fritz Perls in the '60s and admired the way I had integrated Gestalt therapy into my work. He then proceeded to rattle off a long list of woes.

"At school I try to take care of everything, all the details. It's killing me. Now about my personal life—my wife is more suspicious and cautious than me. I'm easygoing and expressive; she's inhibited. I want to connect with her, but . . ."

He went on nonstop. Suddenly I dropped to the floor, pantomiming, particularly with my shoulders, the huge weight of his troubles. Amazed, he stopped and laughed. (Gestures and physical movements can be extremely funny; pantomimists know that.)

"Of all the items on your list, which is the most important right now?" I asked.

"Love," he said, though it had not been part of the list.

"Okay. So now lie down on your back," I instructed. He closed his eyes. When I touched his feet, I was surprised to find them soft and delicate. "Are you a walker?" I asked.

"Not really. Only three miles a day." He was very serious, but I heard the contradiction in a humorous way. Three miles to me means *walking.*

"Let's go to your hip joint." I moved my hands under his left hip, and I found several objects in his back pocket. "Okay," I said with a twinkle, "we've got a 'wallet' release, a 'money' release, a 'key' release and a 'credit card' release." Gregory laughed, began to relax and took a deep breath. I asked him to imagine what his hip joint would say to him.

" 'I'm not talking. I'm staying quiet and I'm not going to let you know what's going on.' " He wriggled around vigorously. (Another obvious contradiction.)

I touched his right hip and asked him what he noticed on that side of his pelvis.

"A big blank," he said emphatically.

"Hey, that's terrific! People meditate for years to be blank." Again he laughed. I placed my hands under his head; his eyes remained closed. I could feel him go more deeply into a trance. "What are you experiencing from my touch?"

"Love, warmth . . . pain and anguish," he said, suddenly sobbing. "Now I feel the pain and anguish move through me. And I feel the warmth, love and support from your hands." He began to wail. "I want support, love and warmth very badly." Without warning, his arms and legs began to flail in all directions.

I asked him to open his eyes and look at me. He did, and after a few moments, his eyes became more focused. He admitted that he felt lonely with his wife because she did not receive his love and warmth. "I want her to *feel* warmth and love the way I do. I'm so sad that she's missing all my love!"

"Perhaps she receives your love in her own way."

"I want her to experience life the way *I* do." His voice trailed off.

"Greg, can you tell your wife she isn't you?"

"Sally, you're not me. You're you!" He repeated this phrase several times. Each time more color flushed his face. Suddenly, he sat up. "Wife, you are not me!" He cast his eyes down and went into his usual "suffering trance" state. "It's difficult for me to accept her the way she is."

I asked him if he remembered Professor Higgins's song from *My Fair Lady.*

His face lit up. "Why can't a woman be more like a man," he said immediately.

"Would you be willing to sing the song?" I asked.

"Yes," he replied without hesitation. "I'm emotional, lovable and warm. Why can't she be more like *me!*" He sang and doubled over, laughing.

"By George, you've got it!" I sang back in a British accent.

"I'm so lovable—why can't she be more like me?" He continued singing and laughing. We all started to laugh. Everyone applauded.

Greg's feeling of isolation maintained his loneliness, distanced him from people and prevented him from making authentic contact with his wife. Humor, sandwiched in between serious slices, was essential to interrupting Greg's "suffering trance" state and helped him to realize that his expectations were unrealistic.

Virginia: "My Vagina Is Laughing"

Virginia is in her forties, a pale woman with almost translucent skin, short-cropped brown hair and dark circles under her eyes. She looks around the room nervously. Once in a while she stops, stares at me and describes her good work in education—the many changes she's made creating an exciting curriculum. Yet she appears to be frightened, cautious and very shy.

She lies on the table. When I touch her head, she closes her eyes, and I get an image of a delicate person. Her neck is stiff. I sense I had better move more slowly and gently. A few times she jerks suddenly when I touch her.

"I don't want to move away when I'm touched," she says in a soft voice.

"When was the last time you were touched and didn't jump?" I ask.

"My older brother held me in his arms and taught me how to whistle. I was three years old." She smiles slightly at the cherished memory.

"Show me how you whistle," I say cheerfully.

She creates a soft, warbling tone, like a bird, then suddenly shifts to a shrill, high-pitched sound, as though she were hailing a cab. The two extremes make everyone in the room laugh.

"Where are you afraid to be touched?" I ask.

"In the genitals. My vagina," she replies quietly.

I understand the seriousness of what she's saying, yet instinct prompts me to use humor. "I want to tell you a precious story that happened with my three-year-old godson, Adam," I say. "He dragged me into the bathroom, pulled down his pants, sat on the toilet and said with a big grin, 'Lana, do you have a 'gina?'

" 'Yes, Adam,' I responded with a straight face.

" 'I don't have a 'gina, do I?'

" 'No, Adam, you don't.' I smiled slightly.

" 'Daddy doesn't have a 'gina. Could you show me *your* 'gina?' He looked at me beseechingly.

" 'No, I don't think so,' I answered, stifling a giggle.

"At this point, Adam and I started to laugh, and I told him he should ask his mom to show him hers."

Virginia's body softens and shifts dramatically as I tell her the story. Still, I know she has to be touched carefully.

"My vagina is laughing," she says, surprised at her own words.

A ripple of laughter starts deep in her belly and travels to the top of her head.

"Can your vagina whistle?" I ask.

She roars with laughter, her entire body shaking. Her eyes are open, and she is out of her trance. Her neck loosens and energy flows freely throughout her body. She rolls over and sits on the edge of the table. Suddenly she looks deeply into my eyes. "*I've* got a laughing vagina!"

She gets up slowly, looks around and makes eye contact with several people. "I still feel shy," she says as she walks up to a group member she trusts and repeats, "I've got a laughing vagina!" Her energy is vibrating.

I didn't yet know the cause of her fear. But humor was necessary to lighten this sensitive subject matter. Telling Adam's story invited Virginia's identification with his natural curiosity and delight.

Creating safety was (and is) fundamental. She appreciated my asking permission to touch her each time I did. The next stage of her development will be her ability to integrate humor about a very serious subject—her sexuality and what made her so fragile.

I trust by now that you see how important humor is in providing us with psychological flexibility. It is not the only tool in the therapist's kit, but used circumspectly, it is a valuable resource and antidote to sadness and fear.

Before I get to the body/mind exercises in the next chapter, I'll close this one with a simple exercise I often use on myself and with my groups. You can practice it by yourself.

EXERCISE:
LAUGHING MEDITATION (SOLO)

Close your eyes and go back to some scene in your life that was funny. Allow a smile to emerge. Travel in your body and notice where you experience the humor. Place your hands on those areas and smile more broadly. Take a few deep breaths and start to laugh out loud by simply saying, "Ha, ha, ha," then change it to "Hee, hee, hee." Soon you'll be laughing freely–not watching yourself, just letting loose. When you finish, slowly open your eyes and look around. Write down whatever you wish in your journal. If you share this exercise with other people, tell them your story and what struck you as so funny.

13.

Mind Your Muscles®: A Seven-Day Body/Mind Exercise Program

You are entitled to be flexible in every aspect of your life. This chapter is devoted to specific ways of stretching, becoming aware and moving tight physical areas that still complain after you have explored the emotional causes for their discomfort.

Guiding Principles and Suggestions

- Each person is built differently physically and emotionally. It's vital to go at your own pace. Each time you repeat the exercises, you'll become more familiar with them and integrate their benefits into your physical and emotional life.
- The more pleasure you have while exercising, the more you will want to continue. The nervous system doesn't learn with pain, but with plea-

sure, so don't push yourself too far. Listen to pain's warning and stop if you hear it.

- Watch out for overzealous stretches. The body protects itself by putting on the brakes, sending an alarm for your muscles to contract. Listen to it.
- Be gentle with yourself. By moving slowly, with awareness, you'll know when your muscles have had enough. Success depends on your ability to listen to and hear what your muscles are saying.
- You're the boss. You can stop anytime, for any reason. There is no need to force any movement.
- Your brain and nervous system learn to relax and become more acute when you are aware of how your body moves. Practicing many of the exercises while lying on a firm surface will give you the sensory feedback necessary for becoming aware and sensing the changes in your being.
- Allow emotional and energetic streams to flow through you while stretching. You might be surprised by sudden anxiety or tears as you perform certain moves. If you need to stop, do so, rest and then continue.
- You may feel confused; it's part of the change process. Have fun with this state rather than becoming anxious. Laughing is helpful.
- Write and/or draw in your journals before and/or after the exercises.
- If you hear a critical voice ("You're not flexible"; "Your grandmother could do these better than you"), send it to Bermuda and have it nag someone else.

The Seven-Day Body/Mind Exercise Program

For the first week, do the following exercises in sequence as I've outlined them, since each day builds on the one before. When you complete the seventh day, you may then mix and match, depending on what your body needs. You'll find that your body will be more flexible, your awareness heightened, your breath deepened—and you'll build up strength as an additional bonus.

The exercises were initially inspired by Moshe Feldenkrai
Technique and my own movement training. They have been r
years of teaching. My particular innovation is to allow and en
as well as physical expression. Be sure you don't push yourself if you're un
fortable. If for any reason you can't do a certain part of an exercise, close your eyes and imagine yourself doing it. It's almost as good as the real thing.

The body/mind exercises are:

1. Arranged for a combination of specific body areas and are done lying on the floor, sitting or standing.
2. Divided into sections, so you can either do one section at a time or the whole exercise.
3. Numbered to guide you step by step through the exercise. The time estimates allow for several repetitions of each step.
4. Prepared with specific rest places built in. Take advantage of these pauses in order to integrate the movements and feelings into your body and mind.
5. Designed so you can do them at any time. Many are done on the floor (keep your office door closed!), and some can be done in a car, plane, waiting on line, sitting at your desk; they are valuable before and after playing a sport. However, they often lead to the best results if you do them in the early morning and/or evening when other events of the day won't intrude. You need no special clothing or equipment.

DAY 1

MOVING LIKE A BAMBOO TREE
(*FOR HEAD, NECK AND UPPER BACK*)

If you've ever experienced a pain in your neck and a stiffness in your shoulders, you're like billions of others around the world. This exercise will help you to listen to what your neck is saying and to relax your shoulder and neck muscles, thereby allowing your head the flexibility to move and see in all directions.

Preparation: Sit comfortably in a chair.

Time: 5 minutes.

Goal: To relax the head and neck.

Added bonus: Softens upper back, stretches rib cage and promotes easeful eye movements.

1. Move away from the back of your chair so you are not leaning against it and place your feet on the floor. Rest your arms on your lap and take a few breaths.

2. Roll your eyes to the right as far as you can and allow your head, neck and torso to follow your eyes without straining. When you cannot move any farther, choose a spot directly in front of your eyes and mentally photograph it.

3. Return to your starting position, facing forward. Move your left hand over the top of your head, covering your right ear. (This is called a "head cradle.") Gently pull your head down to the left with your face still facing forward.

4. As you bend to the left, close your eyes and imagine something flexible in motion—a bamboo tree bending in the wind, for example. Continue bending to the left several times and slowly listen to your ribs . . . neck . . . and spine.

5. Return to your starting position, shake your hand out, rest it for a few moments on your lap. Keep your eyes closed.

6. Move your right hand over the top of your head, cover your left ear in the head cradle position and gently bend to the right with your face still facing forward. Again, imagine scenes that are "soft" and relaxing. Continue bending to the right several more times, pause and return to your starting place. Shake your hands out, rest them on your lap and take a deep breath. Open your eyes.

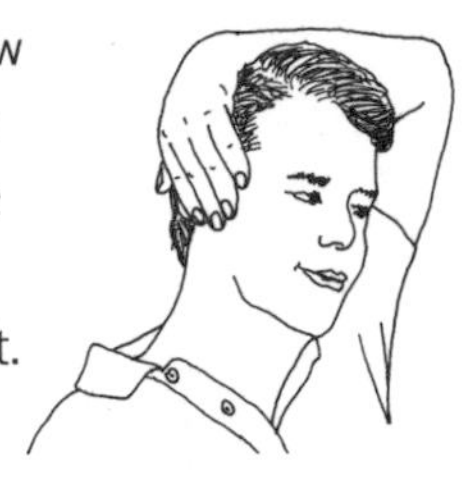

7. Again "cradle" your head with your left hand. Now roll your eyes to the left, with your head, neck and torso following and twisting (as if you are looking at someone behind you). Return to the starting position. Your head, neck and torso will move as one unit, twisting to the left. Pause, return to the starting position (retaining the head cradle) and take a breath.

8. Remain in the head cradle position and again twist your head, neck and torso to the left. However, this time roll your eyes to the right. (Your eyes move in the opposite direction from your body.)

9. Return to the starting position and repeat this movement combination several times, body moving to the left, eyes to the right. Pause, return to the starting position and rest your arms on your lap.

10. Now roll your eyes to the right, as you did at the beginning of the exercise, and continue turning to the right with your head, neck and torso following your eyes. Turn a bit farther. Did you pass the original spot you "photographed" at the beginning? (I'll bet you did!) Congratulations.

11. Return to the starting position. Close your eyes and travel through your body. Do you feel any differences? If so, what are they?

12. Stand up slowly, pause and then walk around. What do you feel right now? Has the exercise changed your posture and walking in any way? What was the most confusing moment?

13. Repeat the entire exercise, beginning by rolling your eyes to the left side. Notice if there are any more changes when you reverse directions.

ROLLING THE HEAD
(*FOR HEAD AND NECK*)

Preparation: Lie on your back on a mat or carpet. Keep a water bottle nearby.

Time: 5 minutes.

Goal: To learn how far your head moves and to experience the degree of tightness of your neck.

Added bonus: Releases eye tensions and may soften tops of shoulders.

1. Stretch your legs out fully, rest your arms at your sides and let the back of your head rest on the floor.

2. Slowly roll your head from side to side. Notice how far your head moves to the right and how far to the left. Return to your starting position.

3. Send your awareness down to your knees. Allow your knees to float up toward the ceiling and place your feet on the floor. Let your lower back release and move down toward the floor. Roll your head to the right, then to the left. Are there any differences? If so, what are they?

THE BOTTLE HEAD ROLL

Preparation: Lie on your back on a mat or carpet. Have an empty bottle nearby.

Time: 7–8 minutes.

Goal: To use your hand to move your head so that your neck is passive.

Added bonus: Relaxes neck muscles, releases eye tensions and softens tops of shoulders.

1. Send your awareness down to your right hand. Place an imaginary or real bottle underneath your right palm. Roll it up (toward your chest) and down (toward your feet). Observe how the fingers, palm and heel of your hand move as you roll the bottle.

We will go through each movement slowly.

2. Feel the bottle underneath your palm. As you roll the bottle downward, it will end up under the heel of your hand. As you roll it upward, it will pass underneath your palm and finish under your fingertips. As you roll it down again, the bottle begins at your fingertips, passes underneath your palm and finishes under the heel of your hand. Your hand is *actively* causing the movement of the bottle, while the bottle is *passive* and *is* moved. We will take this part of the exercise and apply it to moving your head.

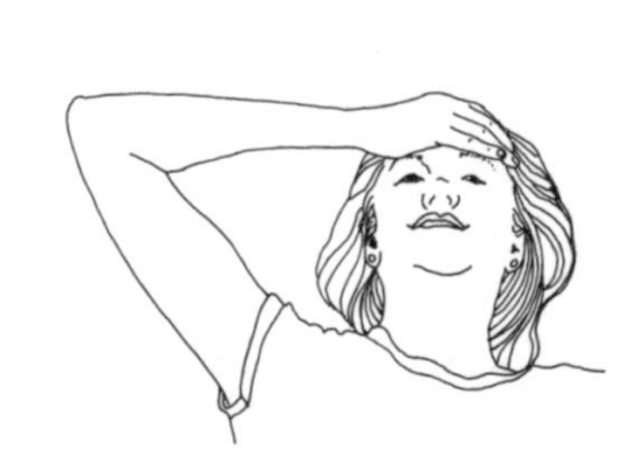

3. Place your right palm across your forehead. The heel will be near your right temple and the fingers near your left temple. Just like the bottle, your hand rolls your head toward the left. When your head turns as much as it can, your fingers and palm will be off your forehead and only the heel of your hand will be touching.

4. Now, use your right hand to roll your head to the right, beginning with the heel and then the palm touching your forehead and finishing with the fingertips. When your head turns as much as it can, the heel and palm will be in the air and just your fingertips will be touching your left temple.

5. Use your right hand to roll your head back to the middle. Move it slowly away from your head, shake it out and rest it at your side.

6. This time use your left hand and repeat Steps 1–5.

7. When you have finished, rest. What do you experience?

IMPORTANT: After a few moments, roll your head from side to side again without your hands. (Going back to the beginning of "Rolling the Head.") Are there any differences now? What are they?

Roll onto your side, pause, then slowly sit up. Now move to standing. Walk around and enjoy the mobility of your head and neck.

CRADLING THE HEAD
(*FOR HEAD,NECK, SHOULDERS AND BACK*)

Preparation: Lie on your back on a floor covered by a mat or carpet.
Time: 7–8 minutes.
Goal: To soften the upper back, chest and neck muscles so that the head, neck and back can move more freely.
Added bonus: Upper back and rib cage stretch; movement sideways promotes spine flexibility.

1. Stretch your legs out, place your arms at your sides and allow the back of your head to rest on the floor. Travel through your body and sense which parts touch the floor and which do not. Roll your head from side to side. Notice how far it moves to the right and to the left.

2. Float your knees up toward the ceiling and place your feet on the floor. Again, roll your head from side to side and notice any differences.

3. Roll over onto your right side (bending your knees at about a 90-degree angle) and rest your left leg on top of the right one. You can support your head with your right arm if you wish.

4. Move your left hand over the top of your head and slide it under your right ear. You are now doing the head cradle on your side.

5. Using your left hand (your head is relaxed and passive), lift your head slowly up toward the ceiling and back down onto the floor. Repeat this movement several times, keeping your eyes focused straight ahead. Notice what happens to your ribs when your head lifts up. (Your left ribs squeeze together and your right ribs stretch.)

6. Move your left hand and arm away from your head, shake them out and roll over onto your back.

7. Float your knees up toward the ceiling, place your feet on the floor and rest your arms at your sides. Close your eyes again.

8. Roll your head from side to side. Take a few moments to listen to any differences between the right and the left sides. What do you experience in your back?

9. Roll over onto your left side. Repeat Steps 3–8 using your right hand for the head cradle.

10. When you have finished, roll onto your side, pause, then slowly sit up. Now move to standing and remain where you are for a few moments. Now walk and look around the room.

DAY 2

LEGS AND FEET LEAD THE WAY
(*FOR BACK*)

"Oh, my aching back" is a refrain in everyone's life. In these exercises for your feet, legs and hip joints, you will discover how to release the tensions in your lower back and free the head and neck.

Preparation: Lie on your back on a firm mat or carpet.
Time: 6–8 minutes.
Goal: To soften and relax the lower back; to mobilize the hip joints.
Added bonus: Stretches in feet, ankles and legs.

1. Stretch your legs out and rest your arms at your sides. Check in with your body, especially your lower back, and notice how it contacts the floor.

2. Float your knees toward the ceiling and place your feet on the floor. Roll your head from side to side and notice how it moves to the right and left.

3. Roll onto your right side, bending your knees at about a 90-degree angle. Rest your left leg on top of the right one. You can support your head with your right arm if you wish.

4. Stretch your left leg down over your bent right leg until your left knee is straight. Let the side of your left foot rest on the floor. We'll call this the "neutral position." The key here is to maintain a straight left leg during this part of the exercise.

5. Without pushing, lift your straight left leg up toward the ceiling. (It does NOT have to go very high.) The important awareness here is the quality and ease of the leg lift rather than the height. Return it back down over the bent leg and rest the side of your left foot on the floor (neutral position).

6. Send your awareness down to your left foot. Slowly stretch your toes in a downward direction (like a ballet dancer on her toes), then relax your foot and return it to the neutral position. Repeat the stretch several times. Notice what happens to your calf muscles, knee and thigh. Rest your left foot back in its neutral position.

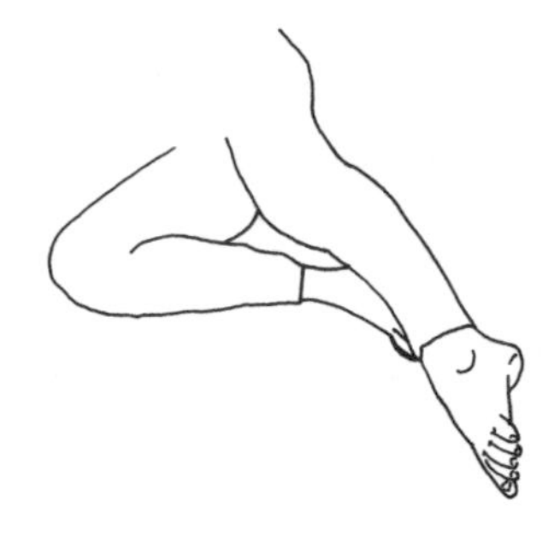

7. Slowly flex your left foot by moving your toes upward as though you were standing on the floor. Relax your left foot and return to the neutral position. Repeat the flex several times and notice what happens to your calf muscles, knee and thigh. Let everything go and rest your left foot in its neutral position. Remember, your left knee is straight and your foot in its neutral position.

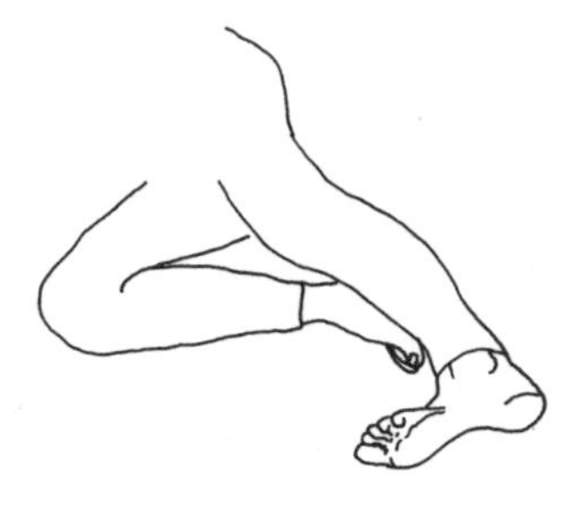

8. Now raise your left leg up and down again. What do you notice about the difference in quality and height of the raised leg now? Is the movement smoother? Easier? What do you feel in your left hip joint? Rest your left leg in its neutral position.

9. Send your awareness to your left foot. Flex it and lift your left leg up and down a few times. What do you notice about the quality of lifting with your foot flexed? What happened to your ankle, calf, thigh and hip joint?

10. Stretch your left foot, pointing the toes downward like a ballet dancer, and raise your left leg up and down several times with a stretched foot. Notice the differences between lifting your leg with a flexed foot and lifting your leg with a stretched foot. What are they? Return to the neutral position with your foot relaxed.

11. Lift your straight left leg up and down again. Do you remember how the "lifting" felt in the beginning? What do you notice about the quality of the leg lift now? Is it more easeful? What do you experience in your left side and in your lower back?

12. Let everything go, roll over on your back, stretch your legs out and rest your arms at your sides. Take a few deep breaths. Listen to how your body is lying on the floor right now and roll your head from side to side. Any differences?

13. Float your knees up to the ceiling, place your feet on the floor. Roll your head from side to side again. What do you feel in your lower back? Could you squeeze your hand underneath it? What differences do you feel between the right and left sides of your body? Rest for a few moments.

14. Roll over onto your left side, bending your knees at about a 90-degree angle. Rest your right leg on top of your left. You can support your head with your left arm if you wish.

15. Now stretch your right leg down over your bent left leg until your right knee is straight. Let the side of your right foot rest on the floor in the neutral position.

16. Repeat the exercise from Steps 5–15.

17. After you have rested for a few moments, roll over onto your side, pause, then slowly sit up. Now, move to standing. Stand for a few moments and sense any changes in the relationship between your feet and the floor. Walk and look around the room.

FLYING LIKE A BIRD
(*UNITING THE UPPER AND THE LOWER BODY*)

You are now ready to combine "Cradling the Head" (Day One) with "Legs and Feet Lead the Way" (Day Two). Although you can do them separately, there is a unique effect on the whole body when you combine them. So here goes!

Preparation: Lie on your back on a firm mat or carpet.

Time: 6–8 minutes.

Goal: To integrate the upper body (head, neck, back and spine) and the lower body (pelvis, lower back, hip joints, legs and feet).

Added bonus: The ribs experience a more intense stretch. The hip joints and pelvis open up farther.

1. Stretch your legs out and rest your arms at your sides. Travel through your body and sense how you are lying on the floor. Experience your upper back, lower back and pelvis.

2. Float your knees up toward the ceiling and place your feet on the floor. Roll your head from side to side and notice how it moves to the right and left.

3. Roll onto your right side, bending your knees at about a 90-degree angle. Rest your left leg on top of the right one. You can support your head with your right arm if you wish.

4. Stretch your straight left leg down over your bent right leg. Let the side of your left foot rest on the floor in the neutral position.

5. Move your left hand over the top of your head and slide it under your right ear in the head cradle.

6. Using your left hand, lift your head up and down a few times, keeping your eyes straight ahead. Pause for a moment and rest your head on the floor or on your right arm. Continue to cradle your head while your upper body rests.

7. Send your awareness down to your left leg and foot. Lift your left leg up and down a few times and then rest it.

8. Now lift BOTH your cradled head and left leg at the same time and suspend them in the air for a few moments. Then float them both back down to the floor. Repeat this movement a few times. Imagine a bird flying in the air or experience any other image that emerges.

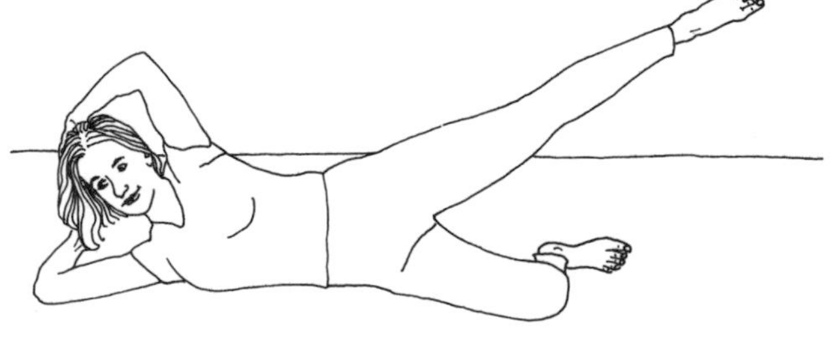

9. When you have finished, let go and roll onto your back. Stretch out your legs, place your arms at your sides and rest. What do you notice about your body now? How does it feel lying against the floor? Roll your head from side to side and notice any differences.

10. Now roll onto your left side and repeat Steps 3–9.

11. When you have completed this simultaneous movement sequence on both sides, roll onto your back, stretch your legs out and rest your arms at your sides. Listen to your body and experience its relationship to the floor. How do you experience yourself now?

12. Roll onto your side, pause, then slowly sit up. Now move to standing. Enjoy walking around your space with a new awareness of your upper and lower body. It's amazing how lower body movements help the upper torso, head and neck to relax. Indeed, every part of us is connected.

DAY 3

ELBOW KISSES THE KNEE
(*FOR LOWER BACK AND SPINE*)

When was the last time you kissed your knees? In the following exercise (and variations) you'll probably begin by "sending" a kiss to your knees. Depending on the flexibility of your spine and lower back, your lips may not actually touch your knees at first.

There are many variations for this exercise; it's a classic! My version is different from the "abdominal crunches" that are taught in many exercise classes. Awareness, stretching and breath are the key elements here, particularly when done slowly.

Preparation: Lie on your back on a firm mat or carpet.

Time: 10–15 minutes.

Goal: To stretch and lengthen the spine; to relax the lower back.

Added bonus: The rib cage and chest become more flexible; the abdominal muscles are strengthened.

1. Stretch your legs out and rest your arms at your sides. Roll your head from side to side. Listen to how your head moves to the left and to the right.

2. Close your eyes and travel through your entire body, especially your lower back. Notice how much space there is between your lower spine and the floor. Slip one of your hands under your lower back and feel that space. Return your hand to your side.

3. Send your awareness down to your knees, float them up toward the ceiling and place your feet on the floor. Using your right and left palms, roll your head from side to side. Shake your hands out.

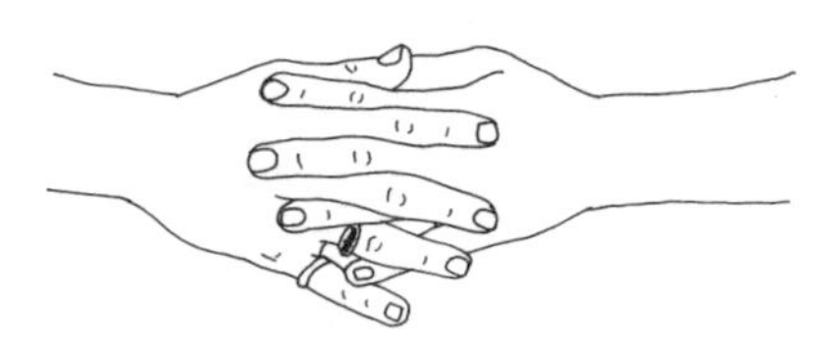

4. Interlace the fingers of both hands and slide them underneath the back of your head. Float both your elbows up slowly and let your interlaced hands lift your head off the floor comfortably. Exhale when you lift your head and inhale when you return it to the floor. Repeat this movement several times, focusing on your stomach muscles and your breathing. Rest.

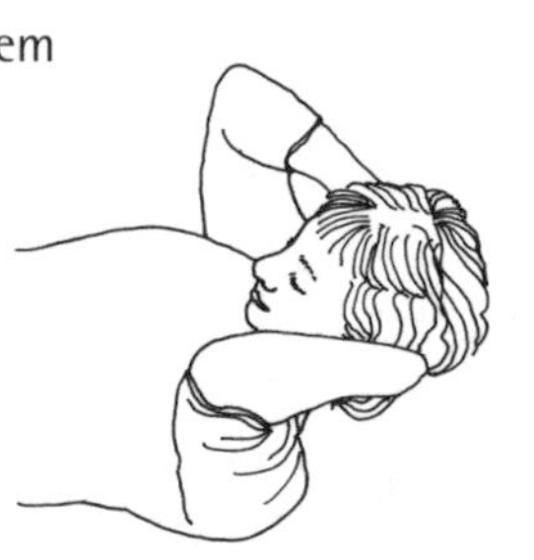

5. Move your knees gently toward your chest, suspend them there at a 90-degree angle and let your feet dangle.

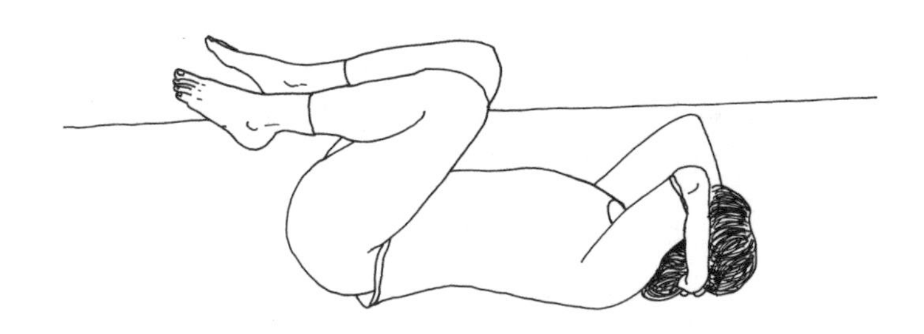

6. Leave your left hand underneath your head and place your right hand around the outside of your right kneecap.

7. Now, open your eyes. Float your left elbow up and let your left hand lift your head. Point your left elbow toward your right knee and send it a kiss. Move your left elbow closer to the knee and send it another kiss. Slowly, let your left hand carry your head back down to the floor. Close your eyes.

8. Repeat this movement several times. Exhale when you send those kisses, inhale when you return to the floor. Enjoy the stretch in your spine and back. When you complete the sequence, place your feet on the floor and rest.

9. Move your knees toward your chest, suspend them at a 90-degree angle and let your feet dangle. Move your right hand underneath your head and place your left hand around the outside of your left kneecap. Repeat Steps 7–8 with your right elbow sending kisses to your left knee.

10. When you complete this movement, shake your hands out and rest them at your sides.

11. Slip one of your hands under your lower back (if you can!) and notice what has happened to the space between your body and the floor. Relish the sensation of having a stretched lower back touch the floor, releasing tension as it flattens out. How do you experience your contact with the floor now? What has changed?

12. Roll onto your side, pause, then slowly sit up. Now move to standing. Walk around and enjoy your lengthened spine and a more relaxed lower back.

VARIATIONS ON THE "ELBOW KISSES THE KNEE" THEME

The following variations are ideal to do immediately following the main theme.

Preparation: Lie on your back on a firm mat or carpet.

Time: An additional 6 minutes.

Goal: To continue softening the rib cage, lengthening the spine and stretching major muscles that affect posture and walking.

Added bonus: Strengthens the abdominal muscles and deepens the breath.

Variation A:

1. Float your knees up toward your chest, suspending them at a 90-degree angle, and let your feet dangle.

2. Leave your left hand underneath your head and place your right hand around the side of your right kneecap.

3. Lower your left foot and place it on the floor, knee bent. Now float your left elbow into the air and let your left hand lift your head. Point your left elbow toward your right knee and send it a big kiss. Slowly, let your left hand guide your head back down to the floor.

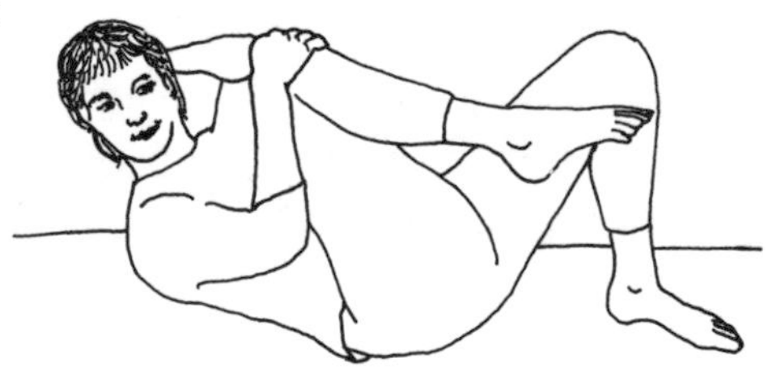

4. Repeat this movement sequence and add a breathing pattern. Exhale when your elbow sends a kiss and inhale when your head returns to the floor. When you have completed several repetitions of the kiss, you are ready for the next variation.

Variation B:

1. Your left hand is under your head and your right hand is placed around the side of your bent right kneecap. Stretch your left leg out onto the floor.

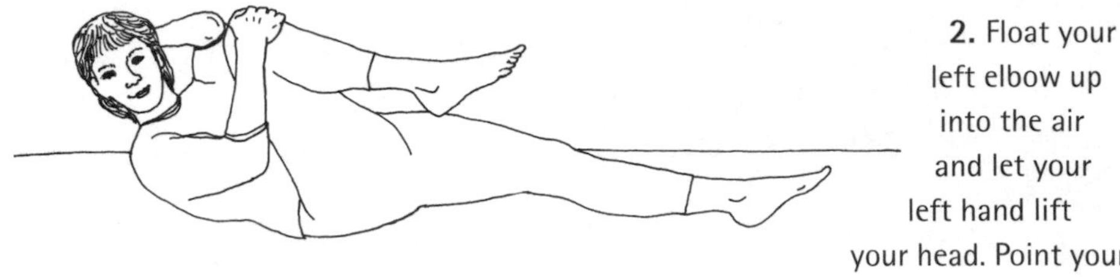

2. Float your left elbow up into the air and let your left hand lift your head. Point your elbow in the direction of your right knee and send it a kiss. (Remember—your left leg remains stretched out on the floor.)

3. Add the breathing pattern. Exhale as you send a kiss to your knee and inhale when your head moves back down to the floor. Exhale as you begin to lift your head again. Repeat several times. Now return to Variation A. Return your left knee up and place your left foot on the floor. With your left hand, lift your head and send kisses to your right knee with your left elbow.

4. When you have completed several repetitions of Variation A, you are ready to return to the original main theme and sing it in an easier way.

5. Float your left knee up toward your chest (both knees are now in the air at 90-degree angles). Your right hand is still around your right kneecap.

6. Let your left hand lift your head and point your left elbow in the direction of your right knee and send it a kiss. Notice how you experience your back as you move this time around. Is it easier for your left elbow to make contact with the right knee? Since your arms probably didn't grow longer, what stretched?

7. Allow both your feet to come down to the floor. Shake your hands out and interlace them once more. Slide them under the back of your head. Let your hands lift your head up and look toward your knees. Do you remember when you first moved this way? Have you come closer to your knees?

8. Move your hands away from the back of your head, shake them out and rest them at your sides. Send your awareness to your lower back again and notice how it feels on the floor. What are you experiencing right now?

9. Roll onto your side, pause, then slowly sit up. Now move to standing. What is your experience standing on the floor at this moment? How does your spine feel? Walk around and thank your elbows for kissing your knees so many times.

DAY 4

REACHING FOR THE SKY
(*FOR SHOULDERS AND SHOULDER BLADES*)

If you habitually stoop and hunch over your work, you will probably develop round shoulders, and your shoulder blades may become frozen in a rigid position. This will constrict your breathing and chest expansion.

Preparation: Lie on your back on a firm mat or carpet.
Time: 6 minutes.
Goal: To melt and move stuck shoulders and shoulder blades.
Added bonus: Softens upper spine; frees head and neck.

1. Rest your arms by your sides, close your eyes and take a few deep breaths.

2. Allow your knees to float toward the sky and place your feet on the earth.

3. Focus on your upper back, shoulders and shoulder blades. Notice how they contact the floor. Are there any differences between the right and left sides of your back?

4. Roll your head from side to side. How much does it move to the left and to the right? Notice any differences in the quality of your head movement.

5. Send your awareness to your right hand. Slowly move it up toward the sky until your right elbow is straight. (Leave your right shoulder blade resting on the floor.) Pause for a moment.

6. Open your right hand and stretch your fingers up to the sky (as though reaching for something or somebody).

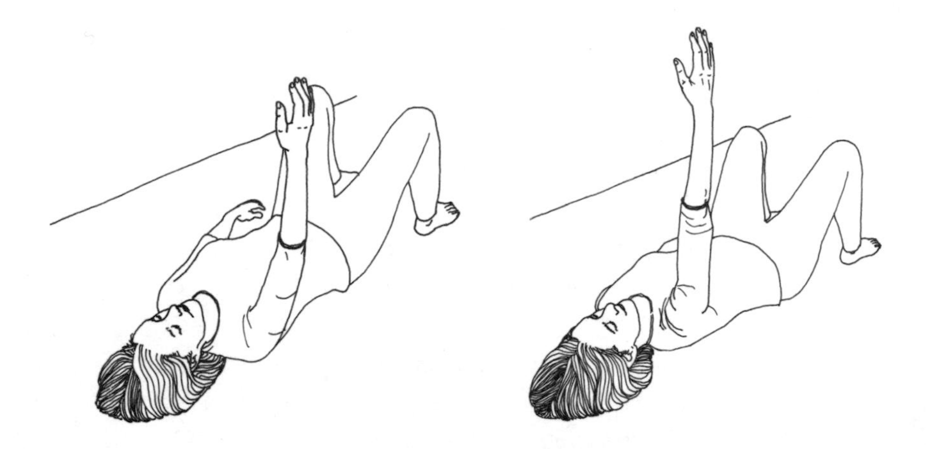

7. Continue reaching upward with your right hand until you feel your right shoulder blade leave the floor. (Be sure to keep your elbow straight.)

8. Remain in this "up" position for a moment and take a deep breath. Sense the stretch between your spine and right shoulder blade.

9. Slowly let your shoulder and shoulder blade move toward the earth. (Your arm is still straight, fingers pointing to the sky.)

10. Repeat moving your straight right arm upward and downward several times.

11. Allow your shoulder and shoulder blade to return downward to resting on the earth.

12. Bending your right elbow, slowly let your right arm move down and rest your right palm on your chest. Do you notice any differences between your right and left shoulders and shoulder blades? Which side feels closer to the earth?

13. Roll your head from side to side and notice any differences in the quality of movement. Stretch your legs out and rest.

14. Float your knees back up toward the sky and place your feet on the earth. Focus on your upper back, shoulders and shoulder blades.

15. Now continue this exercise from Steps 3–14 focusing on your left arm and hand.

16. When you have completed Step 14, bend your left elbow, slowly let your left arm move down and rest your left palm on your chest.

17. Stretch both legs out, take a deep breath and rest.

18. Listen to your shoulders and shoulder blades. What are the differences now compared with when you started? How do you experience yourself right now?

19. Roll your head from side to side again. Do you experience any differences in quality and range of movement?

20. Roll onto your side, pause, then slowly sit up. Now move to standing. How do you experience your shoulders resting on your rib cage? Start walking around the room and notice how your shoulders, shoulder blades and arms feel.

SHOULDER SYMPHONY
(*FOR SHOULDERS AND SHOULDER BLADES*)

Preparation: Lie on your back on a firm mat or carpet. Have a pillow or a towel nearby.

Time: 5–7 minutes.

Goal: To move and soften the shoulders and shoulder blades.

Added bonus: Lengthens upper spine; frees head and neck.

1. Stretch your legs out. Notice which parts of your body are contacting the floor, especially the back of your shoulders and shoulder blades.

2. Roll your head from side to side and sense how far it moves from the center to the right and to the left.

3. Float your knees up toward the ceiling and place your feet on the floor. Roll your head from side to side again and notice if there are any differences.

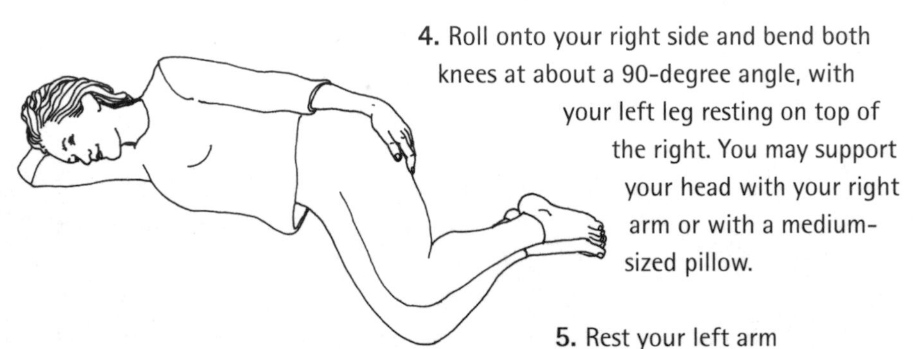

4. Roll onto your right side and bend both knees at about a 90-degree angle, with your left leg resting on top of the right. You may support your head with your right arm or with a medium-sized pillow.

5. Rest your left arm comfortably on your left side with your left palm approximately over your left hip. This position will be called "home base." (You will be returning to home base throughout this exercise.)

6. Sending your awareness to your left shoulder, gently slide it up toward your left ear, then slide it down to home base. Repeat this movement a few times.

7. Starting at home base, slide your left shoulder down toward your left hip, then slide it back up to home base. Repeat this movement several times. Rest.

8. Refocus on your left shoulder and imagine a line connecting your left ear to your left hip.

9. Slide your left shoulder up to your ear and then slide it all the way down toward your left hip. Repeat this sliding motion several times. When you've completed a few repetitions, return to home base.

10. Roll onto your back, stretch out your legs and rest your arms at your sides. Listen to your upper body. What are the differences between your left and your right side? Roll your head from side to side and notice any differences now.

11. Roll onto your right side again. Bend your knees at about a 90-degree angle. Place your left arm on your left side with your left palm approximately on your left hip.

12. Focusing on your left shoulder, slide it forward (in front of you) and slide it back to home base, keeping your legs and pelvis still. Repeat this movement several times, finishing at home base.

13. Slide your left shoulder backward (behind you), then slide it back to home base. Repeat this movement several times, finishing at home base.

14. Now slide your left shoulder softly forward and then slide it back. Repeat this motion several times, finishing at home base. Rest.

15. Imagine a vibrantly colored circle with your shoulder in the middle of it. Slide your left shoulder up toward your left ear, then travel forward on the circumference of this circle downward toward your left hip. Continue around to the back and return to your left ear. (You've completed a full circle motion.)

16. Repeat this circular movement several times. Each time change your focus of attention from the top of your shoulder, to your shoulder blade, to your chest, to your back, to your neck and to your breath. Pause and rest.

17. Reverse the direction of the circle. Slide your left shoulder toward your left ear, then back, down, forward and up again.

18. Repeat this circular movement several times. Each time change your focus of attention from the top of your shoulder, to your shoulder blade, to your chest, to your back, to your neck and to your breath.

19. Roll onto your back, stretch your legs out and rest your arms at your sides. Notice how your body lies on the floor, especially your upper body. This is a special moment for discerning the differences between the two sides of your body. How do you experience them?

20. Float your knees up, place your feet on the floor and roll your head from side to side again. How far does your head move? What is the difference between your two sides? What do you notice about your shoulders, shoulder blades and upper back? What about your chest?

21. Now roll onto your left side and follow Steps 3–19, using your right shoulder for all the sliding movements.

22. When you have finished, roll over on your side, pause, then slowly sit up. Now, move to standing. Allow gravity to perform its mission. Let your shoulders settle down on top of your torso and free your head and neck. Walk around and enjoy all the melodies and harmonies of your body.

DAY 5

TILTING KNEES TO THE EARTH
(*FOR HIP JOINTS AND LOWER BACK*)

Your hip joints and lower back bear stress and weight. Stretching them regularly ensures that they will support you as you walk, run, twist, bend, lift and sit. The following exercises will first focus on your lower body, then invite the upper body to join. The grand finale invites the upper and lower bodies to move together.

Preparation: Lie on your back on a floor covered with a firm mat or carpet.
Time: 8–10 minutes.
Goal: To stretch hip joints and lower back.
Added bonus: Opens pelvis and belly, allowing for deeper breathing.

1. Stretch your legs out and place your arms at your sides. Take a few deep breaths.

2. Notice how the back of your body contacts the earth. Which parts touch the earth and which do not?

3. Send your awareness to your knees, allow them to float up toward the ceiling and place your feet on the earth.

4. Roll your head from side to side. Notice how much it moves from the right side to the left.

5. Lift your right leg off the earth and cross it over your left knee, so that your ankle hangs down over on the left side. This will be the "center position."

6. From the center position, slowly tilt both legs toward the right. Your upper body remains as flat as it can while you tilt your knees. The twist should come from your waist and hips.

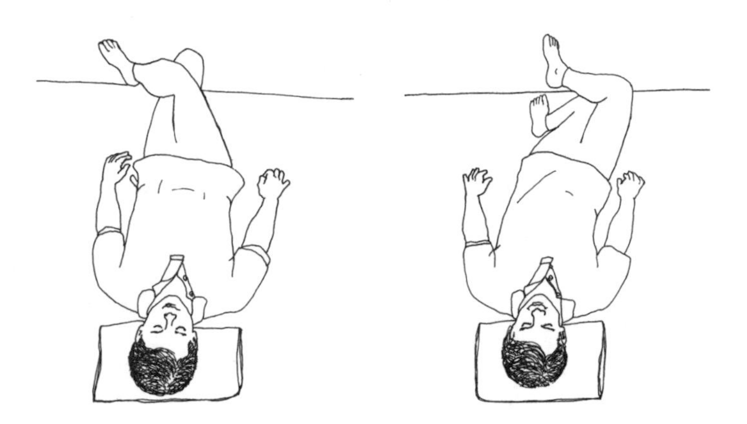

7. Close your eyes and send your awareness to your left hip joint, pelvis and lower back. Which side of your pelvis and buttocks leaves the floor?

8. Return your crossed legs to center position. Repeat tilting both legs toward the right several more times. Notice the distance between your left knee and the floor when you tilt both legs to the right.

9. Let the weight of your tilting legs stay to the right. Place your palms on your belly and breathe in and out rapidly—like a puppy dog. After a few of these puppy breaths, exhale slowly. Move your tilted legs to the center position.

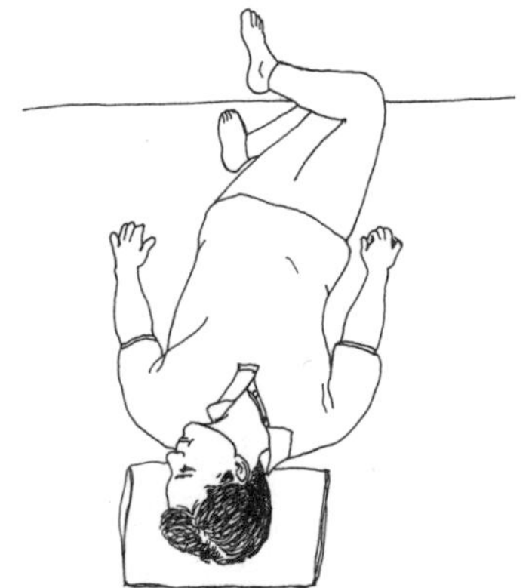

10. Now tilt your legs to the right once more and simultaneously roll your head to the left. Return to center position. Repeat this movement several times (legs to the right, head to the left), finishing in the center position.

11. Uncross your right leg and place your right foot on the floor. Rest. Notice how your body is touching the floor. Do your lower and upper back feel different than they did at the beginning?

12. Your other side has been waiting patiently. So lift your left leg off the earth and cross it over the right knee so that your left ankle hangs down over on the right side.

13. Follow Steps 6–11, this time tilting to the left.

14. When you have finished, roll onto your side, pause, then slowly sit up. Now move to standing. How do you experience yourself standing? Now, walk around and enjoy the movement of your hips and pelvis.

POINTING UP TO HEAVEN
(*FOR SHOULDERS, RIB CAGE AND CHEST*)

Flexibility of your chest, shoulders, shoulder blades and rib cage is crucial for breathing fully, and without good breathing you can't enjoy life's activities to the fullest.

Preparation: Lie on your back on a mat or carpet.
Time: 6–8 minutes.
Goal: To stretch shoulder blades, soften rib cage and expand chest for deeper breathing.
Added bonus: Opens lower back; stretches both sides of upper torso.

1. Stretch your legs out and place your arms at your sides. Breathe normally. Close your eyes.

2. Take a few moments to travel through your body and listen to which parts contact the earth and which do not.

3. Roll your head from side to side and notice how it moves.

4. Send your awareness to your knees; float them up and place your feet two feet apart on the earth. During this exercise, your knees will be continually pointing to the heavens while your upper body moves.

5. Now turn your focus to your hands. Bring your palms together, fingers stretched, and move them up toward the heavens. Let your elbows be straight.

6. With straight arms and palms pressed together, tilt them toward the left. Notice what happens to your right shoulder and your right side as you tilt. Allow your head to follow your arms to the left.

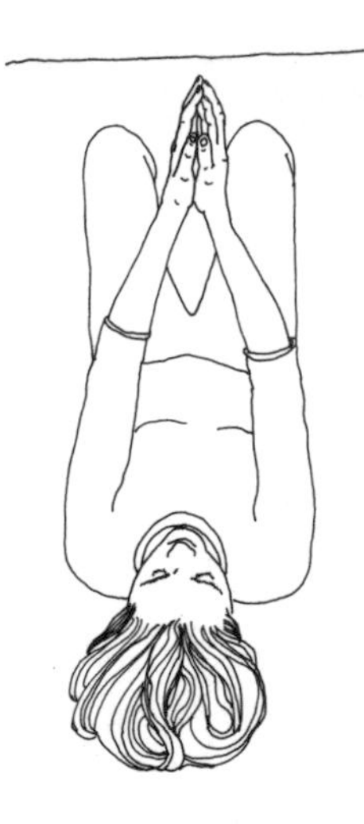

7. Bring your straight arms back to the center position. Rest.

8. Repeat several times, tilting both straight arms to the left and back to center position.

9. When you have finished these repetitions, return to center position, bend your elbows and gently place your palms on your rib cage. Listen to your chest expand and contract as you breathe in and out. Feel its movements with your hands.

10. Roll your head from side to side again and notice any differences in the movement from when you started. How does your body lie on the earth right now?

11. Go back and follow Steps 4–10, this time tilting your arms to the right. Remember to keep your knees pointed up toward heaven.

12. Roll onto your side, pause, then slowly sit up. Now move to standing. How does your upper body feel in this position? Are there any differences now? Walk around and enjoy the space between heaven and earth.

THE GRAND FINALE: BRINGING HEAVEN AND EARTH TOGETHER

Earth and heaven are part of a larger whole: the universe. So it is with your body. The lower and upper parts are part of a larger whole: you!

In this exercise, you will be moving in opposite directions in order to become more centered and balanced (another paradox).

Preparation: Lie on your back on a firm mat or carpet.

Time: 6–8 minutes.

Goal: To connect the upper body to the lower body.

Added bonus: Softens spine as it twists in opposite directions; opens the chest and pelvis to deeper breathing.

1. Stretch your legs out and place your arms at your sides. Take a few breaths.

2. Notice how your body contacts the earth. Which parts touch the earth and which do not?

3. Roll your head from side to side and notice how it moves.

4. Send your awareness to your knees; allow them to float up toward the heavens and place your feet two feet apart on earth.

5. Lift your right leg and cross it over the left knee. In this position, slowly tilt both legs toward the right. (Your upper body remains flat on the earth and your face is looking straight up.) Return your legs to center position.

6. Uncross your legs and place both feet on the earth.

7. Place your palms together (your fingers pointing up) and move your arms with straight elbows up to the heavens.

8. Tilt both straight arms to the left (palms still pressed together) and then return them to center position.

9. Now lift your right leg and cross it over your left knee again.

10. Combining the two previous movements, tilt your crossed legs to the right and simultaneously tilt your straight arms to the left. Then return to center position. Repeat this movement several times.

11. Now switch your legs so that your left leg crosses your right knee. Tilt your legs to the left and your straight arms (palms pressed together) to the right at the same time. Repeat this movement several times.

12. Let everything go. Stretch your legs out and rest your arms at your sides. Roll your head from side to side. Is the movement different from before? Roll onto your side, pause, then slowly sit up. Now move to standing. Imagine the top of your head floating up toward heaven and your feet sinking into the earth. You are connecting the heaven and earth within you. Walk around and enjoy!

DAY 6

WALKING UP THE MOUNTAIN
(*FOR SHOULDERS, SHOULDER BLADES AND NECK*)

People continue to "shoulder" burdens, carrying problems, feelings and tensions in that area. This exercise emphasizes releasing those burdens, feelings and tensions.

Preparation: Stand anywhere.
Time: 10 minutes.
Goal: To stretch, soften and free shoulder blades.
Added bonus: Relaxes neck, head and eyes.

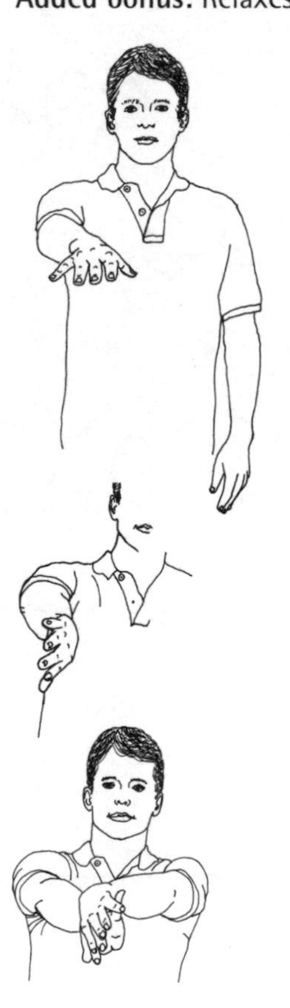

1. Float your right arm up to shoulder height straight in front of the midline of your body. Stretch your fingers forward and separate them. Rotate your arm so that your thumb points down toward the floor.

2. Move your left arm up to shoulder height and cross it over the right arm. Gently interlace the fingers of your hands. Now bring them down together until your knuckles point to the floor. Close your eyes

3. Imagine that you are about to take a walk up a mountain. It is a beautiful day for this trip, so here we go.

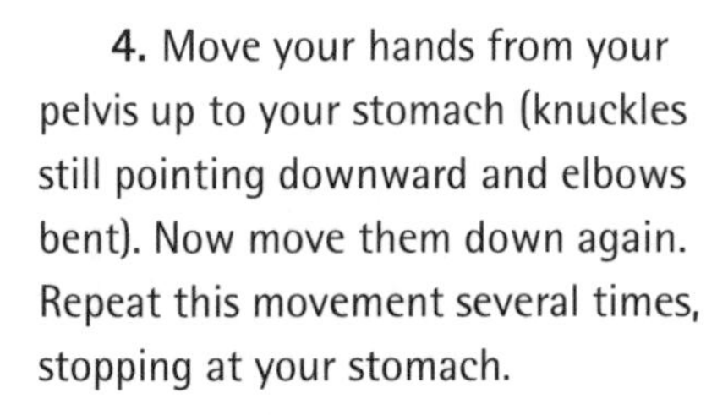

4. Move your hands from your pelvis up to your stomach (knuckles still pointing downward and elbows bent). Now move them down again. Repeat this movement several times, stopping at your stomach.

5. Imagine a gate on this path. Your hands are the key and you open the gate by turning your interlaced hands inward (knuckles pointing at your stomach). Continue to turn them until your fingers point up toward your chin. You have now passed through the gate and you continue to move up the path.

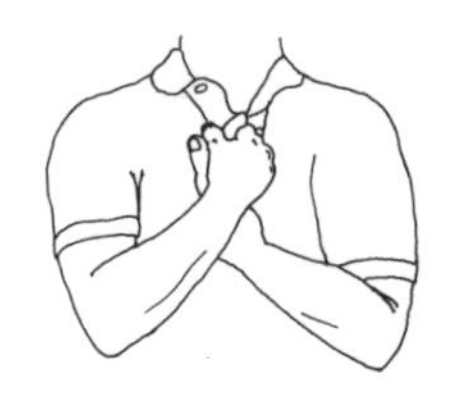

6. Move your knuckles up to your chin and back down to your chest (knuckles still pointing up). Repeat this movement several times. Still imagining that beautiful mountain path, you pause at your chin.

7. Move your interlaced hands (knuckles up) to your nose and back down to your chin several times and stop in front of your nose.

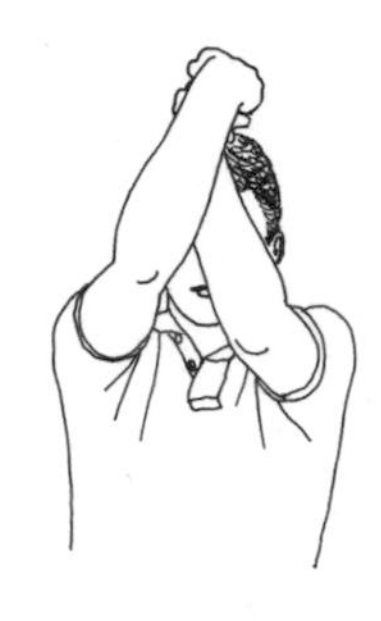

8. Move your knuckles up to your forehead (or near the top of your head) and back down to your nose. You've reached the top of the mountain. Stay there a few moments. Wait. Now begin your descent.

9. Move your interlaced hands (knuckles facing up) to the front of your throat.

10. Open your eyes and look straight ahead. Standing on the mountain ledge, you can see into the distance.

11. Make a large circle around your face with your interlaced hands. Repeat this circle a few times in one direction, then change to the other direction. Complete this activity and rest.

12. Move your interlaced hands (knuckles up) down to your chest.

13. Move your interlaced hands to the left (in a straight line) and turn your head and eyes to the right at the same time. Return your head and hands to the center position.

14. Now move both hands to the right (in a straight line) and turn your head and eyes to the left at the same time.

15. As your interlaced hands are approaching the gate again, they turn inward (knuckles pointing at your chest) and then downward toward the bottom of the mountain. Move your hands all the way to the end of the path.

16. Release your interlaced hands and shake them out. Allow them to relax and rest at your sides. Close your eyes and listen to your shoulders, shoulder blades, neck and breath. What do you notice? Are there any differences between your two sides? Open your eyes and rest. Take a few breaths.

17. Repeat Steps 2–16. However, this time float your left arm to shoulder height. Stretch your left fingers forward, and separate your thumb. Rotate your arm so that your left thumb points down toward the floor. Then cross your right arm over your left and interlace your fingers.

18. When you have finished the entire sequence, slowly begin walking around the room. Allow your shoulders to rest on your torso as you move. I hope you enjoyed your mountain walk.

DAY 7

SOFT EYES
(*FOR EASING YOUR EYES*)

Your eyes work continually and it is easy for them to become tired and stressed. Here are two short yet powerful exercises for relaxing and nourishing your eyes. You can do them anywhere.

Preparation: Lie on your back or sit.
Time: 3–5 minutes or however long you want.
Goal: To relax the muscles around the eyes.
Added bonus: Relaxes head and neck; may improve vision.

1. Close your eyes and take a few breaths. Allow your eyelids to rest comfortably.

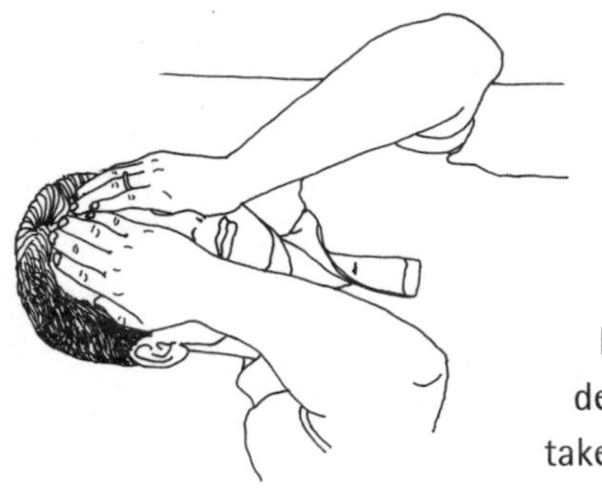

2. Rub your hands together and shake them out. Gently place your palms over your eyes, without pressing onto the eyeballs.

3. Imagine a golden, nourishing, warm light pouring into your eyes. Take several deep breaths. Softly tell your eyes to rest and take in this healing nourishment.

4. Slowly move your hands to the side of your head and into the air.

5. Shake out your hands. With your fingertips, brush the air near and around your eyes. Let your hands brush off any fatigue and negative images.

6. Slowly open your eyes, come back into the room and look around. Are there any differences now?

You can do this exercise several times a day, especially if you've been staring intently at something for a while (say a computer or detailed work).

EYE CIRCLES
(*FOR EASING YOUR EYES*)

Preparation: Sit anywhere comfortably.
Time: 3–5 minutes, or longer if you wish.
Goal: To relax the muscles around the eyes.
Added bonus: Softens face and neck muscles; may improve vision.

1. Close your eyes and take a few breaths.

2. Rub your hands together and gently place them over your eyes. Allow the darkness to surround your eyes—a lovely velvet black.

3. Keeping your eyes closed, slowly roll them to the right and then to the left.

4. Now roll your eyes up toward your forehead. Pause a few moments and roll your eyes downward to your chin. Repeat these movements several times, remembering to breathe as you do them.

Note: You see only a small part of your eyes when you look in the mirror. In fact, each one is about the size of a golf ball. Imagine all the muscles that are needed to stabilize and support them so they can move freely.

5. Do you experience any difference between your right and left eye as you move them inside your head? Shake your hands out.

6. Place your palms gently over your eyes and move them in circles, first to the right, then to the left. What you need to see will come to you.

7. Slowly open your eyelids ("soft eyes"). Come back into the room and look around. Has anything changed? What do you notice? How does your face feel?

This exercise can be done anywhere and at any time you wish. Your eye muscles are as important as any set of muscles in your body.

THE HAND DANCE
(*FOR FINGERS, WRISTS AND FOREARMS*)

Your hands are precious. Holding, reaching, playing music, painting—and, of course, touching—are connections to a rich life.

Computers (or anything involving repetitive hand action) have spawned a new variety of hand and wrist problems. As a pianist, I created several daily hand-wrist-forearm exercises to ease the stress of practicing, but they are effective for many other activities as well. Why wait until your hands hurt when there are stretches and movements you can do to prevent these physical aches and pains. Here are a few:

Preparation: Sit comfortably.
Time: 10 minutes, depending on how much you do.

Goal: To stretch, move and release tensions in the fingers, wrists and forearms.

Added bonus: Softens shoulders, shoulder blades and upper back.

Fingers

1. Place both feet on the ground. Rest your hands in your lap, close your eyes and take a few breaths.

2. Send your awareness to your right hand, pause, and then send it to your left hand. Notice how the two lie in your lap. Are there any differences between them?

3. Float your right hand up slowly until your palm faces your chest. (Check that your elbow and shoulders are relaxed.) Spread your fingers apart (as if opening a fan) for a few moments. Now relax your fingers, letting them form a soft ball. Repeat this movement several times, then rest your right hand in your lap and notice any sensations.

4. Repeat Step 3 with your left hand. When you complete this movement, focus on both your hands and listen to them.

5. Float your right hand up slowly until your palm faces your chest. Focus on your right thumb. Move it in little circles, then enlarge the circle. Repeat this movement several times before relaxing your thumb. Shake your hand out and rest it on your lap. Take a breath.

6. Repeat Step 5 with your left hand.

7. Raise your right hand up again, facing your chest. Open it softly, stretching the fingers into the open-fan position. Move your index finger away from your chest, then toward it several times (your other fingers may move slightly).

8. After you repeat this movement with your index finger, do it with your middle finger, then your ring finger and finally your pinky.

9. Shake your hand out and rest it on your lap. Take a breath. Do you notice any differences between your hands?

10. Repeat Steps 7–9 with your left hand.

Although these are finger movements, you may feel their ripple effect on your wrist and forearm.

Wrists

1. Float your right hand up, palm facing chest. Press your index finger and thumb together firmly, forming a "finger circle." (Other fingers may stay relaxed in a gently curved position.)

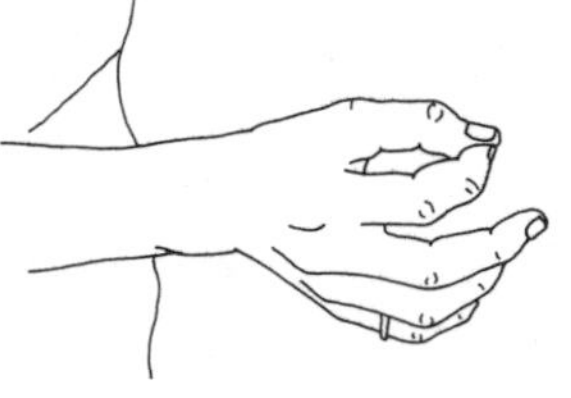

2. Move your right finger circle toward and away from yourself several times. Pause and take a breath.

3. Now move your right finger circle downward several times and then upward several times.

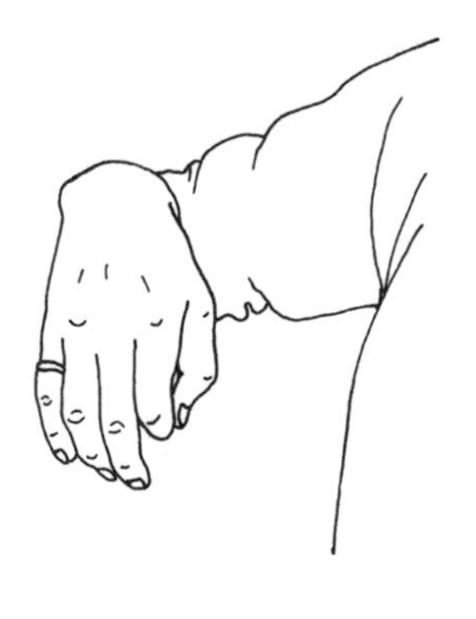

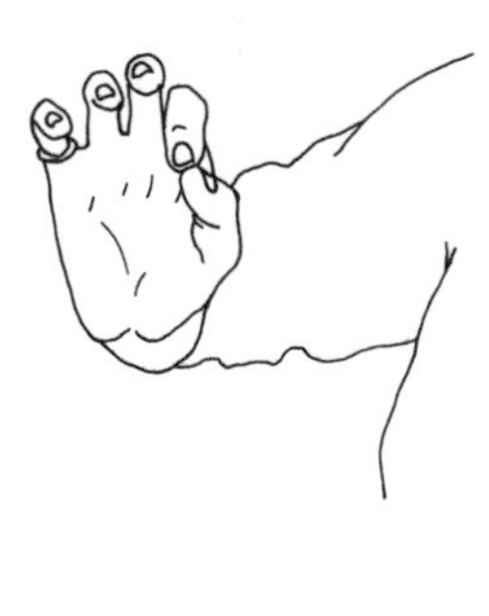

4. Return to the original position (facing your chest), open the finger circle and shake your right hand out.

5. Bring your right hand up again and press your middle finger and your thumb together and follow the instructions of Steps 2–4.

6. Now press your ring finger and thumb together and follow Steps 2–4.

7. Finally press your pinky and thumb together and follow Steps 2–4.

Each time you press a different finger together with your thumb, the physical sensations in your forearm may be different.

8. When you've completed all the finger circles, shake your right hand out and rest it on your lap. Notice any differences between your two hands.

9. Repeat Steps 1–8 with your left hand. Continue breathing throughout as you focus on your hands.

Welcome to the world of flexible fingers and wrists!

Forearms

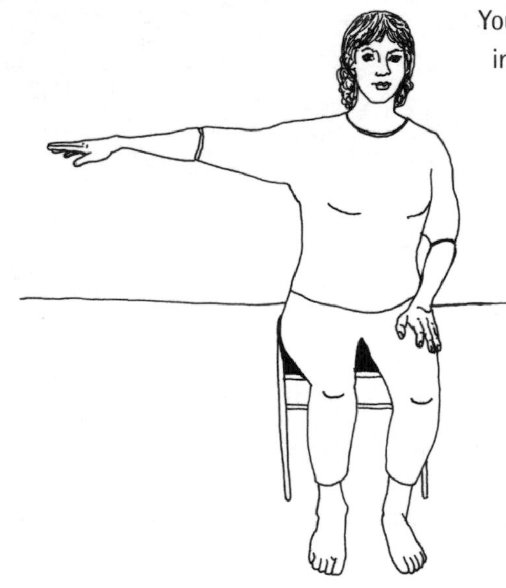

Your fingers, wrists and forearms are intimately connected. Here is a simple yet deep exercise for your forearms:

1. Sitting or standing comfortably, raise your right arm to the side (about shoulder height), leaving your shoulder relaxed.

2. Spread your fingers out softly like a fan. Separate your thumb from the other fingers, your palm facing the floor and your elbow straight.

3. Rotate your straight right arm so that your thumb points upward. Wait a few moments.

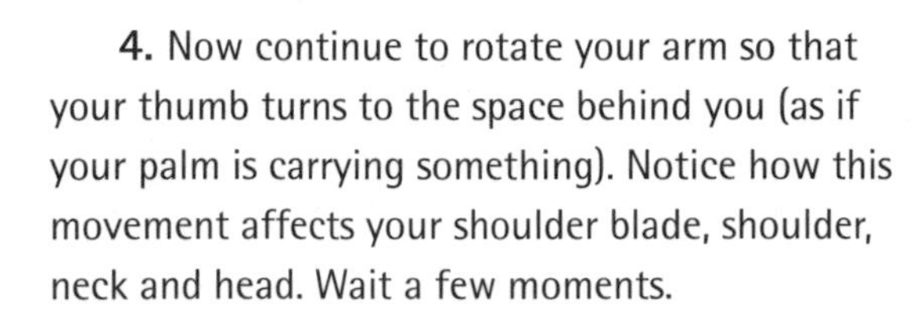

4. Now continue to rotate your arm so that your thumb turns to the space behind you (as if your palm is carrying something). Notice how this movement affects your shoulder blade, shoulder, neck and head. Wait a few moments.

5. Now, rotate your arm so that your thumb points downward. Wait a few moments.

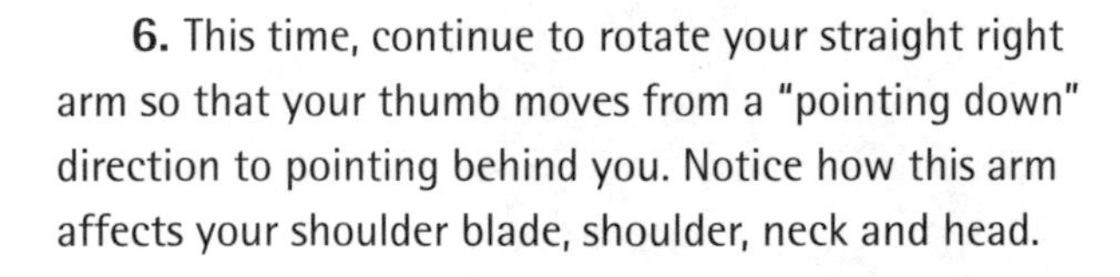

6. This time, continue to rotate your straight right arm so that your thumb moves from a "pointing down" direction to pointing behind you. Notice how this arm affects your shoulder blade, shoulder, neck and head.

7. When you finish rotating your straight arm, let everything go, shake your hands and arms out and rest them in your lap or at your sides. Do you notice any difference between your right and left sides?

8. Repeat Steps 1–6 with your left arm. Don't underestimate this movement; it is quite profound. Move slowly and gently.

9. When you have finished, shake your left arm out and rest it. Do you notice any differences between your two arms, shoulders, shoulder blades, neck and head?

10. Close your eyes. See your fingers, wrists and forearms as vividly as bright sunshine. Appreciate the gifts of your hands.

Your body is your instrument. When all parts are in harmony, you will be able to move in life with ease, pleasure, integration and readiness for the next stage: your Soulquest.

14.

The Integrated Self: Soulquest

People search for meaning and wisdom in life, longing to find their soul.

For me, "soul" refers to the ancient human quest for connectedness to a larger system—from the self, others, history, nature or the universe to the miracle of being alive.

Clients yearning to connect to their souls ask these questions: "What purpose and meaning do I have in my life?" "How can I go beyond my anxieties and fears and live life more fully?" "What gifts do I have to offer others?" "How and when can I trust my heart's song and intuition?"

Soul questions emerge after you have journeyed through your inner jungles, swum into your deepest oceans, moved through internal emotional landscapes and integrated new choices with old behavior. It is my hope that Rubenfeld Synergy has made it possible to ask those questions, to go on that quest.

Solo to Chorus

Soul yearnings are in each of us individually, and all of us collectively. Whoever my clients are, whatever struggles they may be going through, I offer and teach through *who I am.* They want to know *who they are,* and our special relationship is a duet that connects our lives and souls.

Indeed, it is not surprising to find a full orchestra of soul-seekers at workshops. There, we go beyond solo and duet and enter into a community of people who share themselves, listen to each other, develop empathy and care for others as well as themselves.

The universal theme of one vibrates in the community of many, creating a space for tears, laughter, sorrow, anger, pity, play, acceptance, compassion and soul.

Using and experiencing your physical, intellectual and emotional capacities opens the gateways to wisdom and soul-knowledge. Another way to meet and know your soul is in your dreams.

Lois: "I Am Celebrated"

Dreams are windows to the unconscious mind. Every part of the dream is you, including the "split-off" parts of your personality, the disowned ones. You may not hear them in your conscious state—but in a dream they have a voice and ask to be heard.

In a workshop, Lois, a woman in her forties who has experienced many years of Rubenfeld Synergy, comes to the table, lies down and immediately recounts a significant dream. She dives down to the bottom of a deep ocean. There she sees a grotesque monster, asleep next to a sacred pearl-like rock. Lois's mission is to retrieve this rock and carry it up to and above the surface of the water. But as she sees the monster, she wakes up.

I gently touch her head, and it moves easily. When I contact her feet, she places her palms on her belly and explains that there is a purpose to retrieving the sacred rock. Her voice is calm.

I'm still touching her feet when she suddenly grips the sides of the table and trembles fearfully. "You big, fucking son-of-a-bitch punitive monster!" she screams. I cradle her head and reassure her that she will complete her mission, that no one, not even the monster, can stop her. She reaches out, wrestles the rock from the monster, carries it up, breaks through the surface and lifts this pearl-like sacred rock high above her head toward the heavens.

Since her dream represents every part of Lois, including her "monstrous" part, I ask her to role-play the "Sacred Rock" and tell the dream from its point of view.

As the rock, she discovers that she's been at the bottom of the ocean for two thousand years. " 'What's outside of me is not what's inside,' " she says. " 'I am lifted by a determined loving touch and held like a baby.' "

Lois's hands sweep her body and hover over her throat. " 'I don't want to be dropped and be destroyed!' " she cries.

As Lois, she replies, smiling, "We're going to do this together. Both of us."

When they break through the surface, Lois takes several deep breaths. "You know the movie *Roots*?" she asks. " 'When the black man holds his child over his head to the sky?' " adds Lois as the rock. " 'That's how I feel.' "

"It's a celebration of soul!" Lois-as-Lois exclaims, opening her eyes. "Form, strength, will, love and soul all rising to the surface." I nod. She proudly proclaims that there is a connection between her as the rock and her as the diver. "I am celebrated and whole!"

Lois experiences the ecstasy of being alive by reclaiming every positive aspect of herself. She recognizes her essence and welcomes her soul.

The Paradox of Life

When you are able to accept the paradox of life, embracing "both/and" instead of "either/or," you are ready for your Soulquest. Facts may be important in the quest, but unless we "feel" life, in all its paradoxical complexity, it is an intellectual pursuit, not a fully embodied experience.

In the concentration camps of World War II, opposites existed together. Death and torture were ever-present, yet many prisoners maintained faith and hope and continued to find meaning in life. I was fortunate to be able to work with many survivors and their children, and thus be able to see the paradox firsthand.

Rebecca: The Lost Sister

One such person was Rebecca, who came to see me fifteen years ago on the advice of a therapist who had not been able to give her the help she needed. I was struck by the sight of her. She was only in her late twenties, and her face was that of a Raphael Madonna, but she walked with her shoulders scrunched so far forward her head came almost to her chest.

"My name is Rebecca," she said in a voice so low it seemed softer than a whisper.

"Hello, Rebecca. How can I help you?"

"I'm not sure." Again her voice was muted—I soon learned that she was *incapable* of speaking more loudly.

"Lie down on the table," I instructed her. "We'll see if we can find out where to start."

I had never felt a more rigid body. All her parts seemed made of concrete, except for her hands and wrists, which were virtually without muscle, the appendages of a baby. I touched her head and the back of her neck as gently as I could. Immediately, she began to softly sob.

"Are you sad?" I asked her.

"I don't know." She continued to cry.

"How do you see yourself?"

"I'm invisible," she said in a soft, high-pitched whisper.

"You sound like a child. Can you see yourself as a child?"

"No."

"Why?"

"I'm dead."

"Dead? You're with me here, alive as can be."

"I'm my sister. She's dead."

I knew from her family history, passed on to me by her previous therapist, that she was an only child, born in New York to Polish-Jewish parents who had come to the city after World War II.

"She's dead," Rebecca repeated. "I took her place."

And so we began. I worked with Rebecca several times a week, and gradually her story emerged. She had repressed her history, events which had been told to her but which she unconsciously denied, experiencing them only as the trauma that incapacitated her.

It turned out that she *did* have a sister, also named Rebecca. Her mother, her father, an uncle and the baby girl had been hidden from the Nazis by a Catholic family in a farmhouse outside Vilna. They lived underneath the house in a crawl space, which was so shallow that the adults could not stand without stooping. The Catholic family gave them their food through a trapdoor; for nearly four months they did not go out of their cramped quarters.

The Nazis came regularly, searching the countryside for Jews. One day, trapped in their crawl space, the family heard them overhead, their boots loud against the floor. The baby began to cry. Knowing they would be killed if the soldiers heard the baby cry, the uncle clamped his hand over little Rebecca's mouth. At last, the soldiers left. The family was safe. But baby Rebecca had been suffocated.

And now a new Rebecca was suffocated in her stead.

Both touch and talk were pivotal in uncovering Rebecca's trauma. She had been reliving her sister's story and was paralyzed by it, squeezing herself emotionally and physically into a girl hiding from unspeakable terror. Together, we were ultimately successful in bringing her to an awareness of those forces which had molded her into the dysfunctional person she was. With awareness came change. With change came integration. With integration came wholeness and soul. She was able to let her sister go and begin to experience life through her own self.

Compassion, empathy and presence are the hallmarks of any relationship—especially a therapeutic one—when accompanying people struggling through their life journey. Compassion is the core and heart of my intention. And compassionate touch brings you closer to your essence. Most of us don't have stories as traumatic and profound as Rebecca's. But at some point in our lives, most of us are held back, or even incapacitated, by emotional, psychological and physical blocks.

Those blocks can be removed. It takes compassionate touch and "listening" talk.

Coming Home to My Essence

Everyone I've touched is part of my life's path, and I am part of theirs. We're all connected, no matter from where we've come.

In the 1960s, the world-renowned cellist and conductor Pablo Casals became an inspirational mentor for me. "Look at your bodies—what a wonder they are," he used to say to the group of us who apprenticed with him. "You must cherish one another. You must work to make this world worthy of its children."

Music was my emotional and spiritual awakening, and Rubenfeld Synergy is my Soulquest. It crosses all boundaries and prejudices, teaching me to connect with others without judgment, to touch people's heart place and find peace.

Leading a Rubenfeld Synergy session is like conducting Haydn's *Creation.* It is a journey of great depth. We enter the chaos before creation, hear the birth of the universe, rejoice in the theme and variations of life and discover and integrate the harmonious soul voice that is within each of us. The cello and its music was Casals's soul work, and Rubenfeld Synergy is my soul song.

What is yours? If you peeled away every role you've ever played in life, what would be your essence, your core of life, your song of the soul?

A Guide to Body/Mind Exercises for the Aches and Pains That Demand Attention

When you complete the "Mind Your Muscles" exercise program, you will feel more alive, flexible, and aware. However, there are days (for all of us) when you may experience stiffness, aches, pains, and soreness in particular areas of your body. They cry out and demand extra attention and care. Here is a guide to the body/mind exercises arranged according to the areas they benefit.

Head, Neck and Upper Back

Shoulders, Shoulder Blades, Chest and Upper Back

Lower Back and Spine

Hip Joints and Pelvis

Combining the Upper and Lower Body

Feet and Ankles

Eyes

Hands and Forearms

Resources

For information about:

- The four-year Professional Certification Training Program (which meets three times per year)
- Upcoming workshops, conferences, and other appearances
- Audio- and videotapes
- Professional associations
- Certified workshop leaders

Contact:

The Rubenfeld Synergy Center
Phone: 001-800-747-6897
Fax: 001-212-315-3626
E-mail: rubenfeld@aol.com

In Europe
Laura Steckler, PhD, CRS
Tel: 0044 (0) 1309 690 251
E-mail: lsteck@dircon.co.uk

(When e-mailing, please include your regular mailing address and phone number.)

Visit our websites at:
www.rubenfeldsynergy.com
www.ilanarubenfeld.com
www.thelisteninghand.com

ACKNOWLEDGMENTS

I offer my deepest thanks and unconditional appreciation to collaborator Richard Marek, who tuned in to my voice and helped me to build this book, chapter by chapter. He urged, cajoled, and demanded that I stretch my ability to write. After many revisions, he taught me how to stop.

Words cannot express my heartfelt gratitude to David Peretz, who guided me through emotionally turbulent waters throughout the writing of this book, until I arrived safely on the shore of completion.

I am especially grateful to and wish to acknowledge my editor, Toni Burbank, at Bantam Books. She continually challenged me, supported me, and showed me how to refine what I was saying so that the book would flow and sing like music. She is a literary midwife extraordinaire.

"It would be an honor to represent you," she said with bright eyes and a warm smile. Janis Vallely is both an extraordinary agent and a human being who believes in the importance of healing ourselves and the universe. She continues to shepherd me through many publishing challenges. I am most grateful to her.

I wish to thank all the wonderful people at Bantam Books who were so excited about this book: Ryan Stellabotte, Assistant Editor; Glen Edelstein, Senior Design Manager; Nana Greller, Publicity Director; and Tom Cherwin, copy editor.

A deeply heartfelt appreciation and love to Joan Borysenko, who wrote the masterful Foreword to this book.

I am grateful to Anne Kent Rush for skillfully and beautifully rendering all the illustrations. I wish to thank Bill Miller for photographing the four Certified Rubenfeld Synergists who participated in all of the exercises: Rose Andrejewski, Elly Huber, Lyn Krigbaum, and David Lovins.

I am very pleased to be working with Nancy Kahan, whose public relations know-how and warm personality make dealing with the media as smooth as possible. I thank Nancy Levin, my manager, who enthusiastically continues assisting me in transforming ideas into reality.

Many thanks to Stephen Rechtschaffen, co-founder of the Omega Institute in Rhinebeck, New York, who looked directly into my eyes and said, "Enough already! After all these years, you must write a book, and I know who can help you." He challenged me, and his generosity became a beacon of light.

I honor and pay tribute to my teachers and mentors. They are a part of my history and lineage. Thank you, Judith Leibowitz, Fritz and Laura Perls, Moshe Feldenkrais, Charlotte Selver, Virginia Satir, Pablo Casals, George Nassberg, and Buckminster Fuller. Without you, this book could not have evolved.

My appreciation goes to all workshop participants. You shared your lives with me, and your stories became the fertile seeds that blossomed into this book.

I was fortunate to have "an angel" in the person of Lotte Spicer. She continually put papers away and made sure the chaos did not overwhelm me. I give thanks to Nancy Purnell for locating and organizing material for the book, and to Ray Barker for assisting her.

I deeply appreciate Barbara Litwinka, the Rubenfeld Synergy Center's financial director, for her solid support, going beyond her job to assist me and becoming a good friend. I also thank her brother, Richard Litwinka, for assisting her.

Many thanks and acknowledgments go to Vicki Evans, Gail Slevin, and Ellen Blaney for transcribing sessions, lectures, and exercises and for typing so many revisions of the manuscript.

I wish to thank Adrienne Friedland, my makeup artist, and Xenia Galietti, my hair stylist, for their advice, presence, and gentle hands. Also Sigrid Estrada, photographer, who guided me masterfully through the challenging cover photo shoot.

I give thanks to my team of caregivers, who nurtured me with love and attention, especially Nicole Jansen, a brilliant healer; Marilyn Pappert, an outstanding massage therapist; Mary Ann Smith, acupuncturist and energy healer; and Craig Rubenstein, compassionate chiropractor.

Many thanks to authors and editors who invited me to write chapters and articles. They opened the gate to writing this book: Nancy Allison, Lynette Bassman, Christine Caldwell, Henry Grayson and Clem Lowe, Thomas Hanna and Eleanor Criswell Hanna, Richie Herink and Edwin C. Nevis.

A special appreciation to Richard Simon, the editor of *The Family Therapy Networker*, who respected my method and wrote a remarkable interview in the September/October 1997 issue, and to Laura Markowitz, senior editor, who has written articles about body-oriented psychotherapy and has included the Rubenfeld Synergy Method. Thank you, Sylvia Plachy, for your sensitive photographs that accompanied the interview.

I am grateful to Suzanne Forman, a Certified Rubenfeld Synergist, who wrote a splendid lead article about Rubenfeld Synergy for *Massage Magazine* in the January/February 1998 issue.

I am especially proud of Vicki Mechner, a Certified Rubenfeld Synergist, who gathered and edited forty stories from Rubenfeld Synergy graduates, clients, and trainees and published *Healing Journeys—The Power of Rubenfeld Synergy* in 1998.

I am grateful to my close circle of friends and to my professional colleagues who gave me support, comfort, and encouragement. Thank you to Nancy Mayer, Crystal Hawk, Niela Miller, Barbara Lidsky, Pearl Broder, Susan Falk, Seena Russell, George Barenholtz, Jean Houston, Bob Masters, George Lichter, Anna Halpern, Julie Friedeberger, Barker Herr, and Maggie Cottrell.

I wish to thank my longtime friends and colleagues who had faith in me and my work throughout all these years. Thank you to Will Schutz, Betty Fuller, Belleruth Naparstek, Steven Shafarman, Mia Segal, Deborah Caplan, Judith Stransky, Frank Ottiwell, Sarni Ogus, Rupa Cousins, Marian Woodman, Mel Buchholtz, Gerry Jampolsky, Don Campbell, Larry and Eda

Leshan, Candace Pert, Stella Resnick, Jim Kepner, Don Schwartz, John Pierrakos, Joseph Heller, Judith Skutch Whitson, Don Hanlon Johnson, Judyth Weaver, Larry Dossey, Chungliang Al Huang, Emilie Conrad, David Gershon, Gail Straub, Robert and Judith Gass, Ed Rosenfeld, Gabrielle Roth, Jacquelyn Small, Roger Woolger, Robert Bosnak, Christiane Northrup, Al and Diane Pesso, Dawna Markova, Karl Pribram, Steve Andreas, Joan Halifax, Rabbi Joseph Gelberman, Caroline Myss, Olympia Dukakis, Sonia Nevis, Sheila Collins, Ladonna Smith, Alan and Teri Litman, Kathryn Janus, Maureen O'Hara, Irv Katz, Carlos Warter, Carole Hyatt, Jackie Reinach, Hal Bennett, Pierre Weill, Marty Klein, and Randi Brenowitz.

I thank my new circle of friends and colleagues who have generously supported my efforts to write. Thank you to Peter Einstein, Ann Armstrong, Thomas Claire, Harville Hendrix, Richard Gonzalez, Jon Kabat-Zinn, Bessel A. van der Kolk, Amy Zerner, Monte Farber, Josef and Ann Kottler, George and Helen Kaufman, Joan Oliver, Don and Amba Stapleton, Judyth Ashfar, Olivia Mellan, Grace Gawler, Marcia Emery, and Jeffrey Mishlove.

I thank Jack Knight and Certified Rubenfeld Synergist Tony Reilley for introducing me to their business community.

A special thanks to Judy Collins and Lewis Nelson for their continued unswerving belief in me and in this book.

In the late 1960s, the Esalen Institute in Big Sur, California, invited me to give my first workshops. Thank you, Dick Price, Michael Murphy, George Leonard, Peter Frieberg, Kasia Zajac, and especially Nancy Lunney Wheeler, the director of programming, for having confidence in my leadership and work. Abundant thanks to Certified Rubenfeld Synergists Sharon Keane, Tanzy Mixfield, Margaret Tucker, and Dawn Garcia for assisting me at the Esalen Institute these past years.

I am grateful to Greg Zelonka and Tom Valente, current and former program directors of the Omega Institute; Dinabandhu, president; and Elizabeth Lesser, co-founder, for accepting my ideas and for trusting my leadership and innovations all these years.

When the New York Open Center began, I was there helping to launch its first program. Thanks to Ralph White and Walter Beebe, its cofounders, and Adele Heyman, the program director, for their confidence. Much appreciation to Judi Goldstein, Samuel E. Menaged, Adrienne Ressler, and Judith Rabinor of the Renfrew Center for creating opportunities for Rubenfeld Synergy.

A huge dose of gratitude to the folks at the Association for Humanistic Psychology: M. A. Bjarkman, Rae Baskin, Deb Oberg, Ron Maier, Sandy Friedman, and Georgia Berland. I thank Doug Wilson and Prue Berry of Rowe Conference Center for their humor, love, and confidence.

Many colleagues and friends from Virginia Satir's International Human Learning Resources Network continually reached out to me all along the way. Heartfelt thanks to Hal and Linda Kramer, Bob and Alison Shapiro, Bud and Michelle Baldwin, Yetta Bernhard, Jane Parsons-Fein, Maria Gamori, Bernie and Natasha Brightman, John Vasconsellas, Bunny Duhl, Howard Kahn, Pearl Rutledge, Lori H. Gordon, and Jane Gerber. Also, from the American Academy of Psychotherapy, Sol and Bernice Rosenberg, Irv Bailen, Irma Lee Shepherd, and Joan Fagan.

I thank all Certified Rubenfeld Synergists and present trainees for sharing your tears, laughter, confusion, confidence, sadness, joy, healing, and heartfelt love. I especially appreciate the students of the first training program (in 1977) who persisted in nagging me until I agreed to train them.

I would not have completed this book without the commitment, devotion, and creative teaching of my faculty members in the fourteenth and fifteenth training programs. Much gratitude to *Joe Weldon, *Noel Wight, Joan Wade-Lomas, *Elaine Chapline, *Rob Bauer, *Peggy Shaw Rosato, *Alreta Turner, Dana La Rose, Cappi Lang, and adjunct faculty *Florence Korzinski and Barbara Kent.

My cup is full of gratitude and respect for longtime Certified Rubenfeld Synergy colleagues who served on faculties of earlier programs: *Bernie Coyne, *Judy Swallow, *Millie Grenough, Joan Laken, Susan Falk, Susan Powers, Jayne Gumpel, Wener Kundig, Marjorie Paleshi, Joel Ziff, and Irle Goldman. (* designates faculty who are also members of the Master Synergy Council.)

In addition to the faculty, I applaud all the present Teaching Interns who give fully of themselves: Donna Marie Berry, Gay Marcontell, Tanzy Maxfield, Toni Luisa Rivera, Ruthy Goldfarb, Michael Herman, Lyn Krigbaum, Charlene Lane, Maddy Macdougal, Bill Miller, Patricia Orphenides, Jodi Peppil, and Carol Seewald.

I thank the Certified Rubenfeld Synergists in Canada for organizing workshops there: Marjorie Paleshi, Patti Allen, Valerie Bain, Giselle Robert, Heather Davis, Ken Beal, and K. Sass Anderson.

Special thanks go to Esther Gally for beautifully translating my articles into Spanish.

I am grateful to the members of the Rubenfeld Synergy Standards of Practice Task Force for their commitment to our professions: Mary Robinson, Elaine Chapline, Diane Montgomery-Logan, Peggy Shaw Rosato, Heather Davis, Alreta Turner, and Billy Thompson. I thank Jeanne Reock for her organizational support.

I appreciate the wise counsel from attorneys Geoffrey Menin, Robert Levine, and Dan Gluck before and during the writing of this book.

I honor my sister and friend, Lydia Yohay, a sensitive healer in her own right, who has gifted me with a nephew, Joseph, and a niece, Leora.

My thanks and appreciation to Margaret Tucker, a Certified Rubenfeld Synergist, for asking me to be the godmother of her daughter, Lyric. They remind me to play, have fun, and live life in the moment.

Index

About the Author

ILANA RUBENFELD currently teaches at the Esalen and Omega Institutes and the New York Open Center. She has also been on the faculties of the New York University Graduate School of Social Work and the New School for Social Research. She was awarded the Pathfinder Award by the Association of Humanistic Psychology for outstanding and innovative contributions to the field of humanistic psychology. A brilliant presenter who is known for her humor and compassion, she gives lectures and workshops worldwide and conducts a four-year professional certification training program.